Public Goods, Public Gains

Public Goods, Public Gains

Calculating the Social Benefits of Public R&D

ALBERT N. LINK

AND

JOHN T. SCOTT

OXFORD
UNIVERSITY PRESS
2011

OXFORD
UNIVERSITY PRESS

Oxford University Press, Inc., publishes works that further
Oxford University's objective of excellence
in research, scholarship, and education.

Oxford New York
Auckland Cape Town Dar es Salaam Hong Kong Karachi
Kuala Lumpur Madrid Melbourne Mexico City Nairobi
New Delhi Shanghai Taipei Toronto

With offices in
Argentina Austria Brazil Chile Czech Republic France Greece
Guatemala Hungary Italy Japan Poland Portugal Singapore
South Korea Switzerland Thailand Turkey Ukraine Vietnam

Published by Oxford University Press, Inc.
198 Madison Avenue, New York, New York 10016
www.oup.com

Library of Congress Cataloging-in-Publication Data

Link, Albert N.
Public goods, public gains: calculating the social benefits of public
R&D/Albert N. Link and John T. Scott.
p. cm.
Includes bibliographical references and index.
ISBN 978-0-19-972968-5 (cloth)
1. Technological innovations—Economic aspects. 2. Research,
Industrial—Economic aspects. 3. Cost effectiveness. 4. Public goods.
I. Scott, John T., 1947- II. Title.
HC79.T4L563 2011
338.'064—dc22 2010014235

For Carol and Nancy

Contents

Acknowledgments

Our sincere thanks to the many individuals who were directly and indirectly involved in the writing and publication of this book. We greatly appreciate the confidence of Terry Vaughn, executive editor at Oxford University Press, in our ability to address such a provocative subject as the social benefits of public R&D and to deliver a treatise that treats this topic from both a theoretical and applied economics perspective. We also thank the many individuals at the National Institute of Standards and Technology who assisted us in formulating and completing the case studies presented herein.

Of course, we are grateful for the support of our wives, Carol and Nancy, throughout this project; their support has been of great value and indeed necessary for the project's success.

About the Authors

Albert N. Link is professor of economics at the University of North Carolina at Greensboro. He received a B.S. degree in mathematics from the University of Richmond and a Ph.D. degree in economics from Tulane University. His research focuses on innovation policy, academic entrepreneurship, and the economics of R&D. He is the editor-in-chief of the *Journal of Technology Transfer*. Professor Link's most recent books include *Government as Entrepreneur* (Oxford University Press, 2009), *Cyber Security: Economic Strategies and Public Policy Alternatives* (Edward Elgar, 2008), and *Entrepreneurship, Innovation, and Technological Change* (Oxford University Press, 2007). His other research has appeared in such journals as the *American Economic Review*, the *Journal of Political Economy*, the *Review of Economics and Statistics*, *Economica*, and *Research Policy*. Much of Professor Link's research has been supported by funding organizations such as the National Science Foundation, the Organization for Economic Co-operation and Development (OECD), the World Bank, and the science and technology ministries in several developed nations. Currently, Professor Link is serving as the vice-chairperson of the Innovation and Competitiveness Policies Committee of the United Nation's Economic Commission for Europe (UNECE).

John T. Scott is professor of economics at Dartmouth College. He received a Ph.D. in economics from Harvard University and an A.B. in economics and English from the University of North Carolina at Chapel Hill. His research is in the areas of industrial organization and the economics of technological change. He has served as the president of the Industrial Organization Society and on the editorial boards of the *International Journal of Industrial Organization*, the *Review of Industrial Organization*, and *The Journal of Industrial Economics*. He has also served as an economist at the Board of Governors of the Federal Reserve System and at the Federal Trade Commission. Professor Scott's books include *Market Structure and Technological Change* (Harwood Academic, 1987), *Purposive Diversification and Economic Performance* (Cambridge University Press, 1993 and 2005), and *Environmental Research and Development: US Industrial Research, The Clean Air Act and Environmental Damage* (Edward Elgar, 2003). His research in academic journals has spanned four different decades, and his research about the economics of technological change has been supported by the National Institute of Standards and Technology, the National Science Foundation, the National Academy of Sciences, the World Bank, the United Nations Development Programme, and the Organization for Economic Co-operation and Development.

1

Introduction

In this book, we discuss the systematic application of alternative evaluation methods to estimate the social benefits of publicly financed research and development (R&D). The evaluation methods discussed and illustrated are both economics- and non–economics-based. Nonetheless, our discussion should not be interpreted as a broad overview of all approaches to the evaluation of the social benefits of publicly financed R&D. The evaluation landscape is filled with institutional-specific approaches that have evolved over time (and often have in-house champions); and the academic and policy literatures contain examples of myriad approaches, each with redeeming characteristics. We believe that our treatise on economics-based methods herein systematically and theoretically grounds the practice of program evaluation and thus might serve as a unifying template for others to consider.

After summarizing our views on government's role in the innovation process, we review historical perspectives on public accountability, selected systematic approaches to technology-based program evaluation, and related evaluation metrics. We also illustrate the systematic approaches to program evaluation with four detailed case studies.

Why should public institutions be evaluated? To answer such a basic question, one should consider the broader issue of accountability, namely, should public institutions be accountable for their actions? If the answer is yes, as we believe, as do many U.S. policy makers, then the question of how to evaluate a public institution—technology based or otherwise—becomes relevant, if not mandatory.

The concept of fiscal accountability in the United States is rooted in the fundamental principles of representation of the people and by the people. However, as a more modern concept, accountability can be traced to the Budget and Accounting Act of 1921, and that Act began the modern tradition of fiscal accountability in U.S. public institutions.

Building on the general concept of accountability established in the more recent legislation, including the Government Performance and Results Act (GPRA) of 1993, the focus changed to performance accountability to improve the confidence of the American people in the capability of the federal government, initiate program performance reform, and improve federal program effectiveness and public accountability.

More recent legislation has again emphasized fiscal accountability, especially in the light of the financial crisis of the past several years, although a primary emphasis of the American Recovery and Reinvestment Act (ARRA) of 2009 is job creation and growth.

It is inevitable that managers in any public institution will become advocates for their own research agendas. Watching results on a day-to-day basis and witnessing, for example, the benefits of research and scientific inquiry to which one is committed understandably leads managers, and other research participants, to the intuitive conclusion that their activities are valuable. Rightly or wrongly, their conclusion may not be easily communicated to others, much less quantified in a meaningful way. Thus, when political and administrative superiors ask: But how do you know your organization's research or technology-based investigation is effective? managers often find themselves either dissembling or simply telling success stories.

Aside from national defense, many successful government activities are overlooked by the media, especially those for which the social benefits are esoteric or difficult to explain. What typically gains the attention of the popular press are the more obvious failures of government, such as the Katrina relief effort, although the ARRA of 2009 may modify that focus over time. One reason for this general bias is that methods and metrics associated with measuring the successes of government are subtle and often misunderstood.

Herein we discuss accountability from a U.S. historical perspective as well as from an applied perspective. We illustrate in this book that a clear, precise response to the question of performance accountability is possible through the systematic application of evaluation methods, grounded in economic theory, to document value. We do this with an emphasis on R&D-based and technology-based programs to set the stage for the case studies in the latter chapters.[1] However, our methodologies, being grounded in economic theory, are certainly generalizable.

It is important to distinguish between program assessment and program evaluation, regardless of whether the program is R&D or technology based. Although many use the terms interchangeably with reference to public-sector activity, and we note this in Chapter 3, we believe that a distinction is warranted. Policy assessment is based primarily on the criterion of effectiveness: Has the program met its stated goals and objectives; have its designated outputs been achieved? Program evaluation is based on the criterion of efficiency: How do the social benefits associated with the program compare to the social costs? Critical to the program evaluation methods discussed in this book is who conducts the R&D under question: Is the publicly funded R&D publicly performed or is it privately performed? We explain that the answer and economic theory together imply the appropriate evaluation method.

In Chapter 2, we discuss government's role in innovation from the perspective of economic theory. In Chapter 3, we review public accountability issues from both a constitutional and an historical perspective. In Chapter 4, we present systematic approaches to program evaluation—one approach is non-economics and three approaches are economics based, and we describe three widely accepted evaluation metrics typically used. The program evaluation approaches and related metrics are then illustrated in the four detailed case studies that follow. Chapter 5 introduces the four case studies: the Advanced Technology Program's intramural research awards program is the focus of Chapter 6, wavelength references for optical fiber communications are evaluated in Chapter 7, the Malcolm Baldrige National Quality Award is evaluated in Chapter 8, and technologies for the integration of manufacturing applications are evaluated in Chapter 9. The book ends with a brief concluding statement in Chapter 10.[2]

Notes

1. In addition, this focus reflects our own research interests and agenda.
2. Related topics that are not discussed in this book are as important as the topics we address. Specifically, we do not discuss private-sector R&D investment evaluation methods. Our focus is solely on the evaluation of the output and outcomes traceable to public-sector R&D performed in either the private sector or the public sector.

2

Government's Role in Innovation

2.1. Theoretical Basis for Innovation Policy

One theoretical basis for government's role in market activity is the concept of market failure.[1] Market failure is typically attributed to market power, imperfect information, externalities, and public goods. The explicit application of market failure to justify government's role in innovation—in R&D activity in particular—is a relatively recent phenomenon within public policy.

Many point to President George Bush's 1990 *U.S. Technology Policy* as the nation's first formal domestic technology policy statement. Although it was an important initial policy effort, it failed to articulate a foundation for government's role in innovation and technology. Rather, it implicitly assumed that government had a role, and then set forth the following general statement (Executive Office of the President 1990, p. 2): "The goal of U.S. technology policy is to make the best use of technology in achieving the national goals of improved quality of life for all Americans, continued economic growth, and national security."

President William Clinton took a major step forward from this 1990 policy statement in his 1994 *Economic Report of the President* by articulating first principles about why government should be involved in the technological process (Council of Economic Advisers 1994, p. 191): "The goal of technology policy is not to substitute the government's judgment for that of private industry in deciding which potential 'winners' to back. Rather, the point is to correct market failure . . . ."[2]

Subsequent Executive Office policy statements have echoed this theme; *Science in the National Interest* (Executive Office of the President 1994) and *Science and Technology: Shaping the Twenty-First Century* (Executive Office of the President 1998) are among the examples. President Clinton's 2000 *Economic Report of the President* (Council of Economic Advisers 2000, p. 99) elaborated upon the concept of market failure as part of U.S. technology policy: "Rather than support technologies that have clear and immediate commercial potential (which would likely be developed by the private sector without government support), government should seek out new technologies that will create benefits with large spillovers to society at large."

Likewise, Martin and Scott (2000, p. 438) observed: "Limited appropriability, financial market failure, external benefits to the production of knowledge, and other factors suggest that strict reliance on a market system will result in underinvestment in innovation, relative to the socially desirable level. This creates a *prima facie* case in favor of public intervention to promote innovative activity."

2.2. Underinvestment in Research and Development

Market failure, as we address it, could be termed *technological or innovation market failure*. Market failure refers to the market—including both the R&D-investing producers of a technology and the users of the technology—underinvesting, from society's standpoint, in a particular technology or technology application. Such underinvestment occurs because conditions exist that prevent organizations from fully realizing or appropriating the benefits created by their investments. The following explanation of market failure and the reasons for market failure follows closely Arrow's (1962, p. 609) seminal work in which he identified three sources of market failure related to knowledge-based innovative activity—"indivisibilities, inappropriability, and uncertainty."[3]

To elaborate on the concept of market failure, consider a marketable technology to be produced through an R&D process where conditions prevent the R&D-investing firm from fully appropriating the benefits from technological advancement. Other firms in the market, or in related markets, will realize some of the benefits (economic profits—revenues in excess of the opportunity costs of the resources used by these other firms) from the innovation, and of course consumers will typically place a higher value on a product than the price paid for it. The R&D-investing firm will then calculate, because of such conditions, that the marginal benefits it can receive from a unit investment in such R&D will be less than could be earned in the absence of the conditions reducing the appropriated benefits of R&D below their potential, namely the full social benefits. Thus, the R&D-investing firm might underinvest in R&D, relative to what it would have chosen as its investment in the absence of the conditions. Stated differently, the R&D-investing firm might determine that its private rate of return is less than its private hurdle rate; therefore, it would not undertake socially valuable R&D.

This basic concept can be illustrated with Figure 2.1, which follows from Tassey (1997) and Jaffe (1998). The social rate of return is measured on the vertical axis along with society's hurdle rate on investments in R&D. The private rate of return is measured on the horizontal axis along with the private hurdle rate on R&D. A 45-degree line (dashed) is imposed on the figure under the assumption that the social rate of return from an R&D investment will at least equal the private rate of return from the same investment. Two separate R&D projects are labeled as project A and project B. Each is shown, for illustrative purposes only, with the same social rate of return.

For project A, the private rate of return is less than the private hurdle rate because of barriers to innovation and technology (discussed later). As such, the private firm will not choose to invest in project A, although the social benefits from undertaking project A would be substantial (i.e., the social benefits are above the social hurdle rate).

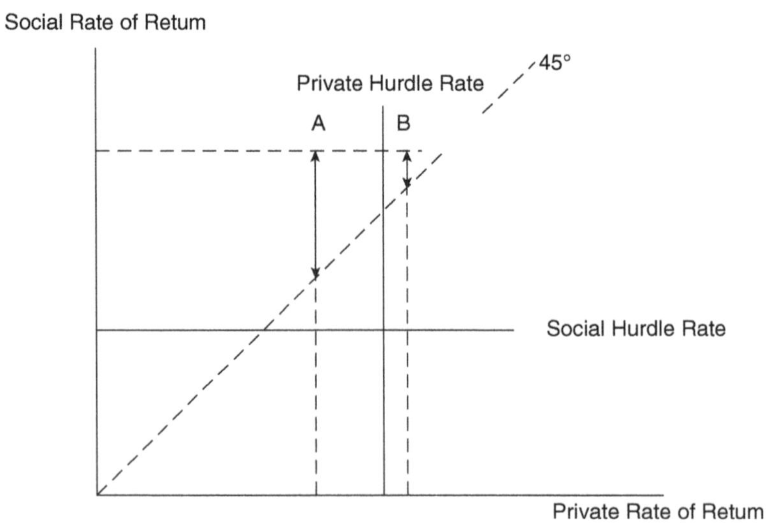

Figure 2.1 Spillover gap between social and private rates of return to R&D.

The principle of market failure illustrated in the figure relates to appropriability of returns to investment. The vertical distance shown between the double arrows for project A is called the spillover gap; it results from the additional value society would receive above what the private firm would receive if project A were undertaken. What the firm would receive is less than its hurdle rate because the firm is unable to appropriate all of the returns that spill over to society. Project A is the type of project in which public resources should be invested to ensure that the project is undertaken. For projects of type A where significant spillovers occur, government's role throughout U.S. history has typically been to provide direct funding or technology infrastructure through public research institutions; this lowers the marginal cost of investment so that the marginal private rate of return exceeds the private hurdle rate.

In comparison, project B yields the same social rate of return as project A, but most of that return can be appropriated by the innovator, and the private rate of return is greater than the private hurdle rate. Hence, project B is one for which the private sector has an incentive to invest on its own, even though the social rate of return is greater than the private rate of return. Or, stated another way, there is no economic justification for public resources being allocated to support project B.

The spillover gap in Figure 2.1 could be thought of as a measure of the consequences—a loss of socially valuable benefits in excess of the opportunity costs—of market failure. Government failure in this context could be evaluated in terms of the effectiveness of public policy in moving project A—shifting its point in the plane of the private and social rates of return—to the point where the socially valuable project is privately profitable.[4] The complete absence of government failure would entail a public subsidy that was just sufficient to shift the point A rightward until the private rate of return equaled the private hurdle rate.

Note that the private hurdle rate is greater than the social hurdle rate in the figure. This is primarily because of management's (and employees') risk aversion and issues related to the availability and cost of capital. These factors represent an additional source of market failure that is related to uncertainty. For example, because most private firms are risk averse (i.e., the penalty from lower than expected returns is weighted more heavily than the benefits from greater than expected returns), they require a higher hurdle rate of return compared to society as a whole that is closer to being risk neutral.[5] Importantly, however, observe that the logic of the market failure is not driven by Figure 2.1's depiction of the social hurdle rate as less than the private hurdle rate.[6] Rather, what drives the market failure is the gap between the social rate of return and the private rate of return. To reduce market failures associated with inappropriability and uncertainty, government typically engages in activities to reduce technical and market risk (both actual and perceived). The following section discusses several circumstances—termed *barriers to innovation and technology*—that cause market failure and an underinvestment in R&D.

2.3. Barriers to Innovation and Technology

There are a number of factors that can explain why a firm will perceive that its expected private rate of return will fall below its hurdle rate, even when the social rate of return exceeds the social hurdle rate.[7] Eight factors are discussed in this section. Individuals will differ not only about a listing of such factors because they are not generally mutually exclusive but also about the relative importance of one factor compared to another in whatever taxonomy is chosen. (See Table 9.1 in Chapter 9 for an example of this fact.) We discuss these eight factors separately in the paragraphs that follow. But individually or in combination, they lead to a private underinvestment in R&D.

First, high technical risk (that is, outcomes might not be technically sufficient to meet needs) could cause market failure. Given that when the firm is successful, the private returns fall short of the social returns, the risk of the activity being undertaken is greater than the firm can accept, although if successful there would be very large benefits to society as a whole. Society would like the investment to be made, but from the perspective of the firm, the present value of expected returns is less than the investment cost and is thus less than the amount yielding its acceptable return on investment.

Second, high technical risk may be augmented by high commercial or market risk (although technically sufficient, the market might not accept the innovation; reasons can include factors listed subsequently such as imitation or competing substitutes or interoperability issues). Given the risk, when the requisite R&D is highly capital intensive, the project could require too much capital for any one firm to be able to justify the outlay. The minimum cost of conducting research is thus viewed as excessive relative to the firm's overall R&D budget, which considers the cost of outside financing and the risk of bankruptcy. In this case, the firm will not make the investment, even when society would be better off if it had, because the project does not appear to be profitable from the firm's private perspective.

Third, many R&D projects are characterized by a lengthy time interval until a commercial product reaches the market. The time expected to complete the R&D and the time until commercialization of the R&D results are long, and the realization of a cash flow from the R&D investment is in the distant future. If a private firm faces greater risk than society does, and as a result requires a greater rate of return and hence applies a higher discount rate than society does, it will value future returns less than society does. Because the private discount rate exceeds the social discount rate, there could be underinvestment. The underinvestment increases as the time to market increases because the difference in the rate is compounded and has a bigger effect on returns further into the future.[8]

Fourth, it is not uncommon for the scope of potential markets to be broader than the scope of the individual firm's market strategies, so the firm will not perceive or project economic benefits from all potential market applications of the technology. Thus, the firm will consider in its investment decisions only those returns that it can appropriate within the boundaries of its market strategies. While the firm may recognize that there are spillover benefits to other markets, and while it could possibly appropriate a portion of them, such benefits are ignored or discounted heavily relative to the discount weight that society would apply. A similar situation arises when the requirements for conducting R&D demand multidisciplinary and multiskilled research teams; unique research facilities not generally available within individual firms; or fusing technologies (i.e., technologies used together or sequentially) from heretofore separate, noninteracting parties. The possibility for opportunistic behavior in such thin markets could make it impossible, at a reasonable cost, for a single firm to share capital assets even if there were not R&D information-sharing difficulties to compound the problem. If society, perhaps through a technology-based public institution, could act as an honest broker to coordinate a cooperative multifirm effort, then the social costs of the multidisciplinary research might be less than the market costs.[9]

Fifth, the evolving nature of markets requires investments in combinations of technologies that, if they existed, would reside in different industries that are not integrated. Because such conditions often transcend the R&D strategies of firms, such investments are not likely to be pursued even if the technologies will be used sequentially rather than at the same time. Underinvestment will occur not only because of the lack of recognition of possible benefit areas or the perceived inability to appropriate whatever results but also because coordinating multiple players in a timely and efficient manner is cumbersome and costly. Again, as with the multidisciplinary and multiskilled research teams, society may be able to use a technology-based public institution to act as an honest broker and reduce costs below those that the market would face.

Sixth, the nature of the technology may make assignment of intellectual property rights difficult. Knowledge and ideas developed by a firm that invests in technology may spill over to other firms during the R&D phase or after the new technology is introduced into the market. If the information creates value for the firms that benefit from the spillovers, then, other things being equal, the innovating firms might underinvest in the technology. Likewise, when competition in the development of new technology is very intense, each firm, knowing that the probability of being the successful innovator is low, might not anticipate sufficient returns to cover costs. Furthermore, even if the firm innovates, intense competition at the application stage can result because of competing substitute goods, regardless of whether they are patented. Especially when the cost of imitation is low, an individual firm will anticipate such competition and might therefore not anticipate returns sufficient to cover its R&D investment costs. Of course, difficulties appropriating returns need not always inhibit R&D investment (Baldwin and Scott 1987). First-mover advantages associated with customer acceptance and demand, as well as increasing returns as markets are penetrated and production expanded, can imply that an innovator wins most (or at least a sufficient portion to support the investment) of the rewards even if it does not appropriate all of the social returns.

Seventh, industry structure might raise the cost of market entry for applications of the technology. The broader market environment in which a new technology will be sold can significantly reduce incentives to invest in its development and commercialization because of what some scholars have called "technological lock-in and path dependency."[10] Many technology-based products are part of larger systems of products. Under such industry structures, if a firm is contemplating investing in the development of a new product but perceives a risk that the product, even if technically successful, will not interface with other products in the system, the additional cost of attaining compatibility or interoperability could reduce the expected rate of return to the point that the project is not undertaken. Similarly, multiple submarkets might evolve, each with its own interface requirements, thereby preventing economies of scale or network externalities from being realized. Again, society, perhaps through a technology-based public institution, could be able to help the market's participants coordinate successful compatibility and interoperability.

Eighth, situations exist where the complexity of a technology makes agreement costly with respect to product performance between buyer and seller. Sharing of the information needed for the exchange and development of technology can render the needed transactions between independent firms in the market prohibitively expensive if the incentives for opportunistic behavior are to be reduced to a reasonable level with what Teece (1980) has termed *obligational contracts*. Teece emphasized that the successful transfer of technology from one firm to another often requires careful teamwork with information-sharing interactions between the seller and the buyer of the technology. In such circumstances, both the seller of the technology and the buyer of the technology are exposed to hazards of opportunism. Sellers, for example, may fear that buyers will capture the know-how too cheaply or use it in unexpected ways. Buyers may worry that the sellers will fail to provide the necessary support to make the technology work in the new environment; or they might worry that after learning about the buyer's operations in sufficient detail to transfer the technology successfully, the seller would back away from the transfer and instead enter the buyer's industry as a technologically sophisticated competitor. Once again, if society can use a technology-based public institution to act as an honest broker, the social costs of sharing technology could be less than market costs.

These eight factors that create barriers to innovation and technology and thus lead to a private underinvestment in R&D are listed in Table 2.1. While we have already discussed these factors individually and have listed them in the table as if they were discrete phenomena, they are, to re-emphasize, interrelated and overlapping, although in principle any one factor could be sufficient to cause a private firm to underinvest in R&D.

2.4. The Role of Public Research Institutions

Public research institutions—their intramural research as well as their focused extramural research activity—could overcome many of the barriers to innovation and technology discussed in the previous section.

Table 2.1 Factors Creating Barriers to Innovation and Technology

1. High technical risk associated with the underlying R&D
2. High capital costs to undertake the underlying R&D with high market risk
3. Long time to complete the R&D and commercialize the resulting technology
4. Underlying R&D spills over to multiple markets and is not appropriable.
5. Market success of the technology depends on technologies in different industries
6. Property rights cannot be assigned to the underlying R&D
7. Resulting technology must be compatible and interoperable with other technologies
8. High risk of opportunistic behavior when sharing information about the technology

We now discuss the role of public research institutions because they are at the heart of three of the case studies presented in subsequent chapters.

To describe the rationale for public research institutions that provide, intramurally or extramurally, infrastructure technology needed by industry, we use a definition of risk that is focused on the operational concern with the downside outcomes for an investment. The shortfalls of the private expected outcomes from society's expected returns reflect appropriability problems. There are several related technological and market factors that will cause private firms to appropriate less return and to face greater risk than society faces. These factors underlie what Arrow (1962) identified as the nonexclusivity and public good characteristics of investments in the creation of knowledge. The private firms' incomplete appropriation of social returns in the context of technical and market risk can make risk in its operational sense unacceptably large for the private firm considering an investment.

Operationally and with reference to Figure 2.1, Tassey (1992, 1997, 2003, 2005), for example, defined risk as the probability that a project's rate of return falls below a required, private rate of return or private hurdle rate (as opposed to simply deviating from an expected return).[11] As illustrated in Link and Scott (2001a), for many socially desirable investments, the private firm faces an unacceptably large probability of a rate of return that falls short of its private hurdle rate. Yet, from society's perspective, the probability of a rate of return that is less than the social hurdle rate is sufficiently small that the project is still worthwhile.

Martin and Scott (2000) made the point that the design of appropriate public policy should match the policy with the specific source of underinvestment. In that light, they identified several roles for public research institutions. Given the types of research they perform, such institutions could be called standards and infrastructure technology institutions. Specific activities of those institutions are matched with specific sources of underinvestment in research, and the various activities are illustrated with examples from case studies.

One role for a public research institution is to facilitate the promulgation and adoption of standards and thereby, for example, reduce the risk associated with standards for new technology as inputs are developed for adopting industries such as in the sectors developing software, equipment, and instruments. We are using the term *standards* in a general sense to refer to voluntary performance protocols and interoperability standards, test methods, and standard reference materials.[12] Although one can find examples where observers have thought that product standards were used in anticompetitive ways, the beneficial role for the promulgation of standards by public research institutions is quite general and important, encompassing several types of standards. The public institution with research capability can respond to industry's needs for standards, working with industry to develop them while serving as an honest broker providing impartial mediation of disputes that could not be provided by a private firm with a proprietary interest in the outcomes.[13] In the absence of the public research institution, industry would have incurred higher costs to replace the public standards activities than the actual costs to the public institution for those activities. Furthermore, the quality of the more costly private standards activities would have been less than the quality of the public standards activities.

For another role, public research institutions can oversee extension services to facilitate technology transfer in sectors such as light industry or agriculture when, for example, small firms, facing limited appropriability from their investments in new technologies yet providing large external benefits to the economy as a whole, apply inputs developed in supplying industries. Such extension services can make possible a vibrant entrepreneurial sector of smaller firms that stimulates the adoption and diffusion of new technology and also innovation, technological advance, and economic growth.

The positive impact of such an entrepreneurial sector has been documented by many scholars (e.g., Audretsch 1995), and in the last two decades recognition of its importance for economic growth has increased and become widespread. Imperfections in credit markets, opportunistic behavior by larger firms that might provide resources to small entrepreneurial businesses, and the unappropriated external benefits from entrepreneurial businesses might require public support of extension services to avoid underinvestment in the transfer of technologies. Although the argument for public research institutions with research capability is not as strong as it is in the foregoing role with standards, public research institutions such as the National Institute of Standards and Technology (NIST) are in a good position to foster the technology transfer provided by extension programs. Such public institutions have knowledge of the key technologies, have working relationships with the industries supplying the technologies, and can assist with the transfer the technologies without opportunistic exploitation of the small firms, allowing them to grow as independent sources of initiative and growth.[14]

For a third role, a public research institute can serve as the coordinator and facilitator for cooperative R&D efforts joining industry, universities, and government in research that is subsidized by the government. The several projects studied by Hall, Link, and Scott (1998, 2001, 2003) provide examples of such cooperative R&D efforts. Such cooperative research with a public research institution as the facilitator is often necessary to coordinate the development of infrastructure technologies as well as precompetitive generic technologies that are at the heart of the development of complex systems involving high cost, risk, and limited appropriability. These complex systems are developed, for example, in aerospace, electrical and electronics, and telecommunications and computer technologies. While the coordination of cooperative efforts that transcend the solely market-based activities of industry is arguably an important and central role for government, the key question is whether a public research institution playing that coordinating role actually needs to have a research capacity itself. Based on the case studies in our other writings, we believe that in many cases the answer is yes.

Finally, for industrial applications of technologies with high science content, where the knowledge base originates outside of the commercial sector, the creators of the knowledge may not recognize the potential applications or effectively communicate the new developments to potential users. As a fourth role in a national innovation system, a public research institution can facilitate the diffusion of advances from research in these cases—such as in biotechnology, chemistry, materials science, and pharmaceuticals—where the applications have high science content. This fourth role is one of facilitating communication and dissemination of ideas from science that can then be used by many sectors to advance applied research and development. In many cases, government funds will have been used by universities to develop the basic science because the ideas have a strong public good component and there would not have been sufficient incentive to develop them without government funding. Once the basic science is available, the knowledgeable public research institution with expertise in both research and connections to industry can help to disseminate the information widely.

Basic research, which has economy-wide implications and very long time horizons, is unlikely to be undertaken by private firms. With this in mind, are there reasons—incentive problems and market failures—that would require the basic research be performed by the government and not by the government's financing of research at universities? That is, are there reasons that the fourth role for the public research institution would include not only working to communicate basic science, transferring it to industry in ways that focus on the industrial usefulness of the basic science, but would include as well the public laboratory actually doing basic science itself? In our experience with the work done in several U.S. government laboratories, we have observed some fairly basic research, yet what is called the most basic of research is in fact quite applied (e.g., using the basic science created in universities to develop new measurement technology). Conceivably there are incentive issues that may dictate the performance of certain types of basic research in the laboratories of public research institutions. By their nature, the research objectives of the government might differ from the interests of universities and their researchers, and it is possible that some goals of the government's basic research agenda would not align well with the current academic interests. Stated differently, academic researchers might find it beneath themselves to do the science that the public happens to want at a particular time. Another possibility is that academic researchers cannot always take the long view—especially given that the long view can change as political administrations change—needed to develop a government-mandated strand of science, in the detail needed, simply because of the constraints of turning out sufficient publications of sufficient variety and quality in the context of review and promotion for the researchers. Laboring in some public service vineyard for a decade or more may not have the necessary academic rewards to ensure survival in the university system. A public laboratory scientist is freed from such constraints and the public laboratory can set its own reward structure that is sensitive to the fact that political administrations and the scientific imperatives of government can change. Finally, national security may dictate that some types of research are performed in government laboratories with heightened security rather than in the more open environment of university laboratories.

In all of the foregoing roles for the public research institution, the institution is an honest broker providing technological services—standards, standard reference materials, calibrations traceable to the standards, technology transfer, and diffusion of scientific advances—without a proprietary, competitive, market-based interest. In addition, the public research institution's research capability is an integral part of developing and maintaining standards and other technological services. The institution is not just an administrator; it is an organization with real scientific and engineering expertise. In matters of generic and infrastructure technology, the institution is an honest broker with leading-edge research capabilities. It also has close working relationships with industry that allow it to understand industry's needs and continually develop and maintain the standards and services that industry relies on for its productivity.

The theoretical foundation for public sector involvement in any aspect of the innovation process emphasizes the government's activities as a way to overcome market failure and create social benefits that the market alone would not provide. That theoretical foundation logically leads to a discussion of public accountability, meaning that the public sector is also responsible for evaluating the social benefits of its actions. We turn to that discussion in the next chapter.

Notes

1. This chapter draws from Link and Scott (2005a) and from the Link and Scott references therein.

2. The conceptual importance of identifying market failure for policy is also emphasized, although without any operational guidance, in Office of Management and Budget (OMB) (1996). The concept of market failure traces to Bator (1958, p. 351) in which he wrote: "What is it we mean by 'market failure'? Typically, at least in allocation theory, we mean the failure of a more or less idealized system of price-market institutions to sustain 'desirable' activities or to stop 'undesirable' activities. The desirability of an activity, in turn, is evaluated relative to the solution values of some explicit or implied maximum-welfare problem." However, aspects of the concept of market failure trace as far back as Sidgwick (1885).

3. Although Arrow (1962) does not elaborate on indivisibilities and inappropriability in his paper, the concepts are well understood in the innovation literature. Recalling that Arrow (1962, p. 609) defined innovation "as the production of knowledge," the market does not price knowledge in discrete bundles and thus because of such indivisibilities market prices may not send appropriate signals for economic units to make marginal decisions correctly.

4. Economists have traditionally emphasized market failure over government failure in their study of markets. Wolf (1979) discussed nonmarket failure

and suggested that it may be equally as important as market failure when studying dynamic markets. For a thorough overview of government failure, see Le Grand (1991) and Winston (2006).

5. There are two parts to the answer to the twin questions of how the social hurdle rate is determined and why it is represented as being less than the private hurdle rate. The first is grounded in the practice of evaluations, and the second is grounded in the theory of public policies to address market failure.

(1) Regarding practice, the Office of Management and Budget (OMB) has mandated that a specified real rate of return be used as the rate for evaluation studies—that is, the rate to be considered the opportunity cost for the use of the public funds in the investment projects. The OMB (1992, p. 9) stated that: "Constant-dollar benefit-cost analyses of proposed investments and regulations should report net present value and other outcomes determined using a real discount rate of 7 percent." That real rate of return (and the related nominal rates derived by accounting for expected inflation rates in various periods of analysis) has been far less than what the respondents in the case studies presented subsequently reported as the private hurdle rate for comparable investment projects in industry during comparable time periods.

(2) Regarding theory, when one evaluates public investment projects, one is invariably looking at cases where there has been some sort of market failure. To improve upon the market solution, the government has become involved (in a variety of ways, in practice) with an investment project. Just as market solutions for the prices of goods may not reflect the social costs for the goods (because of market failure stemming from market power, imperfect information, externalities, or public goods), the private hurdle rates that reflect market solutions for the price of funds—the opportunity cost of funds to the private firms—might not reflect the social cost of the funds. The government might decide that the appropriate social cost—the opportunity cost for the public funds to be invested—differs from the market solution. Typically, in practice, a government could reasonably believe that it faces less risk than the private-sector firms doing similar investments; hence, a lower yield is satisfactory because the public is bearing less risk than the private-sector firm working on its own with a similar investment. More generally, government must decide what the opportunity costs of its public funds will be in various uses, and in general that will not be the same as the market rate. However, all that said, clearly we know from Arrow's (1963) thinking about social choice that the government's decision about what the rate should be cannot possibly reflect the diversity of opinion in the private sector regarding the decision. Consequently, as a logical matter, one could not prove that the government's choice of the right hurdle rate is obviously correct because diversity of opinion about the correct rate will not be reflected in the government's choice.

6. We are evaluating investment decisions—the allocation of capital across alternative investment options, and so the OMB-mandated 7% real discount rate is the one that our evaluations used in the case studies presented later in the book. Commenting on the 7% real discount rate, OMB (2003, p. 33) observed: "The 7 percent [real] rate is an estimate of the average before-tax rate of return to

private capital in the U.S. economy. It is a broad measure that reflects the returns to real estate and small business capital as well as corporate capital. It approximates the opportunity cost of capital, and it is the appropriate discount rate whenever the main effect of a regulation is to displace or alter the use of capital in the private sector. OMB revised *Circular A-94* in 1992 after extensive internal review and public comment." Further, OMB (2003, p. 33) observed: "The pre-tax rates of return better measure society's gains from investment. Since the rates of return on capital are higher in some sectors of the economy than others, the government needs to be sensitive to possible impacts of regulatory policy on capital allocation." However, OMB (2003, p. 33) observed: "The effects of regulation do not always fall exclusively or primarily on the allocation of capital. When regulation primarily and directly affects private consumption (e.g., through higher consumer prices for goods and services), a lower discount rate is appropriate." Hence, if one were evaluating a policy where, instead of alternative uses of investment capital in public R&D investment decisions, the issue evaluated was a regulatory policy (e.g., for health care) that would directly and primarily affect the stream of real income to consumers (e.g., alternative health plans with streams of different magnitudes and different timings), then the OMB has directed (OMB, 2003, pp. 33–34) that "for regulatory analysis" (p. 34), rather than an evaluation of an investment, the real discount rate of 3% should be used and then compared to the results using the 7% real discount rate. OMB explains that for consumers' decisions, 3% better approximates their real rate of time preference.

OMB (2003, p. 33) explicitly stated that the 7% real required rate of return that is based on the average rate of return to private capital investment is "a default position;" yet the market failure story (as captured in Fig. 2.1) recognizes that for investments (not just "regulatory policy") the social rate of return and the private rate of return can (and are expected to) diverge, with the social required rate of return being less than the private hurdle rate. As it turns out, in practice, a 7% social hurdle rate for public investments is not inconsistent with that logic, because the 7% is based on the average, but for the R&D investment projects we evaluated in the case studies the firms report higher private hurdle rates. OMB appears to be taking the least controversial approach by using for the social hurdle rate for investments an average return for private capital investments and by advising consideration of the variance in private returns in different activities. Clearly, as we have noted, there is no reason society should be constrained to its assessments of value by prices determined in markets where there are market failures and the prices give the wrong signals. Hence, the private rate of return on investment should not be expected to equal the social opportunity cost of investment funds; the private rates of return may be based on prices that do not reflect social value. We know that with positive externalities such as unappropriated spillovers that benefit those who did not invest, social rates of return can be high when private rates of return are low. Moreover, the private rate of return can be high even when the social rate of return is low or even negative. For example, in the context of R&D investment, the results of a privately profitable R&D investment may simply cannibalize previously existing economic surplus, causing the investment to have a negative social rate of return. OMB's approach is a solution in the absence of a practical way to determine what the theoretical social hurdle rate should be in any given situation.

7. As Arrow (1962) explained, investments in knowledge entail risk of two types: technical and market. The technical outputs and market outcomes resulting

from technology may be very poor, or perhaps considerably better than the expected outcome. Thus, a firm is justifiably concerned about the risk that its R&D investment will fail, technically or for any other reason. Or, if technically successful, the R&D investment output may not pass the market test for profitability. Furthermore, the firm's private expected return typically falls short of the expected social return as previously discussed. This concept of downside risk is elaborated upon in Chapter 9 and in Link and Scott (2001a).

8. The present value of a cash flow x received t periods in the future given the discount rate r is $x/(1 + r)^t$. So, for example, if the private discount rate is 10% and the social discount rate is 5%, the ratio of private value to social value is about 0.95 for a cash flow in 1 year, but is just 0.39 for a cash flow in 20 years.

9. See Leyden and Link (1999) on the role of a federal laboratory as an honest broker.

10. See David (1987) for detailed development of the ideas of path dependency in the context of business strategies and public policy toward innovation and diffusion of new technologies.

11. Tassey (1992, 1997, 2003, 2005) developed and applied the idea of barriers to innovation and technology using the idea to advance knowledge about appropriate technology policy for the National Institute of Standards and Technology (NIST).

12. Voluntary standards are discussed later with reference to NIST.

13. Industry's scientists and engineers frequently interact with scientists in the public research institution in conferences and workshops, and together they enable the public research laboratories to develop the standards needed as the technological requirements for industry to remain competitive evolve. See the several examples described in Link and Scott (2004b).

14. Private organizations with some public funding have evidently been successful in transferring technology to smaller businesses. Although coordinated by a public research institution, there is substantial private funding for the Malcolm Baldrige National Quality Award Program through NIST (see Chapter 8). The pattern of shared funding among government and private organizations is common to many of the activities of public research institutions—most prominently activities largely performed by the private sector with oversight from the public institution and with some partial public funding of the projects. See Link and Scott (2001b, 2006).

3

Historical Perspectives on Public Accountability

3.1. Historical Origins

Fundamental to public support of economic activity is the public sector's awareness of its accountability for its use of public resources.[1] The concept of public accountability can be traced in the United States as far back as President Woodrow Wilson's reforms, and in particular to the Budget and Accounting Act of 1921. This Act of June 10, 1921 not only required the President to transmit to Congress a detailed budget on the first day of each regular session, but it also established the General Accounting Office (GAO) to settle and adjust all government accounts. We note this fiscal accountability origin because the GAO has had a significant role in the evolution of accountability-related legislation during the past decade.

What follows is a brief review of the legislative history of recent initiatives that fall broadly under the rubric of public accountability.[2] As Collins (1997) noted:

> As public attention has increasingly focused on improving the performance and accountability of Federal programs, bipartisan efforts in Congress and the White House have produced new legislative mandates for management reform. These laws and the associated Administration and Congressional policies call for a multifaceted approach—including the provision of better financial and performance information for managers, Congress, and the public and the adoption of integrated processes for planning, management, and assessment of results. (p. 7)

Fundamental to any evaluation of resource allocation to any program is the recognition that the institution allocating the resources or administering the program is accountable to the public—that is, to taxpayers—for its activities. With regard to technology-based institutions, this accountability refers to being able to document and evaluate research performance using metrics that are meaningful to the institutions' stakeholders—that is, to the public.

3.2. Performance Accountability

3.2.1. Chief Financial Officers Act of 1990

The GAO has a long-standing interest and a well-documented history of efforts to improve governmental agency management through performance measurement. For example, in February 1985 the GAO issued a report entitled *Managing the Cost of Government—Building an Effective Financial Management Structure,* which emphasized the importance of systematically measuring performance as a key area to ensure a well-developed financial management structure (General Accounting Office 1985).

On November 15, 1990, the 101st Congress passed the Chief Financial Officers Act of 1990. As stated in the legislation as background for this Act: "The Federal Government is in great need of fundamental reform in financial management requirements and practices as financial management systems are obsolete and inefficient, and do not provide complete, consistent, reliable, and timely information." The stated purposes of the Act are as follows:

1. Bring more effective general and financial management practices to the Federal Government through statutory provisions, which would establish in the Office of Management and Budget a Deputy Director for Management, establish an Office of Federal Financial Management headed by a Controller, and designate a Chief Financial Officer in each executive department and in each major executive agency in the Federal Government.
2. Provide for improvement, in each agency of the Federal Government, of systems of accounting, financial management, and internal controls to assure the issuance of reliable financial information and to deter fraud, waste, and abuse of Government resources.

3. Provide for the production of complete, reliable, timely, and consistent financial information for use by the executive branch of the Government and the Congress in the financing, management, and evaluation of Federal programs.

The key phrase in these stated purposes, in our opinion, is in point (3) above, ". . . evaluation of Federal programs." Toward this end, the Act calls for the establishment of agency Chief Financial Officers, where agency is defined to include each of the federal departments. And, the agency Chief Financial Officer shall, among other things, "develop and maintain an integrated agency accounting and financial management system, including financial reporting and internal controls," which, among other things, "provides for the systematic measurement of performance."

While the Act does outline the many fiscal responsibilities of agency Chief Financial Officers, and their associated auditing process, the Act's only clarification of "evaluation of Federal programs" is in the phrase "systematic measurement of performance." However, neither a definition of "performance" nor guidance on "systematic measurement" was provided in the Act. Nonetheless, these are the seeds for the growth of attention to performance accountability.

3.2.2. Government Performance and Results Act of 1993

Legislative history is clear that the Government Performance and Results Act (GPRA) of 1993 builds upon the February 1985 GAO report and the Chief Financial Officers Act of 1990. The 103rd Congress stated in the August 3, 1993 legislation that it found, based on over a year of committee study, the following:

1. Waste and inefficiency in Federal programs undermine the confidence of the American people in the Government and reduce the Federal Government's ability to address adequately vital public needs.
2. Federal managers are seriously disadvantaged in their efforts to improve program efficiency and effectiveness because of insufficient articulation of program goals and inadequate information on program performance.
3. Congressional policy making, spending decisions, and program oversight are seriously handicapped by insufficient attention to program performance and results.

Accordingly, the purposes of GPRA are as follows:

1. Improve the confidence of the American people in the capability of the Federal Government, by systematically holding Federal agencies accountable for achieving program results
2. Initiate program performance reform with a series of pilot projects in setting program goals, measuring program performance against those goals, and reporting publicly on their progress
3. Improve Federal program effectiveness and public accountability by promoting a new focus on results, service quality, and customer satisfaction
4. Help Federal managers improve service delivery, by requiring that they plan for meeting program objectives and by providing them with information about program results and service quality
5. Improve Congressional decision making by providing more objective information on achieving statutory objectives, and on the relative effectiveness and efficiency of Federal programs and spending
6. Improve internal management of the Federal Government

The Act requires that the head of each agency submit to the Director of the Office of Management and Budget (OMB) "no later than September 30, 1997 . . . a strategic plan for program activities. Such plan shall contain . . . a description of the program evaluations used in establishing or revising general goals and objectives, with a schedule for future program evaluations." And, quite appropriately, the Act defined program evaluation to mean "an assessment, through objective measurement and systematic analysis, of the manner and extent to which federal programs achieve intended objectives."[3] In addition, each agency is required to "prepare an annual performance plan [beginning with fiscal year 1999] covering each program activity set forth in the budget of such agency. Such plan shall . . . establish performance indicators to be used in measuring or assessing the relevant outputs, service levels, and outcomes of each program activity," where "performance indicator means a particular value or characteristic used to measure output or outcome."

Many speculated, at the time of the passage of GPRA, that it will encourage agencies to ignore what is difficult to measure, no matter how relevant.[4] Alternatively, one could wear a more pessimistic hat and state that GPRA will encourage agencies to emphasize what is easy to measure, no matter how irrelevant.

GPRA was at the time of its passage (Executive Office of the President 2003, pp. 48–49) "the most significant advance in bringing accountability to government programs . . . . Unfortunately, the implementation of this law has fallen short of its authors' hopes. Agency plans are plagued by performance measures that are meaningless, vague, too numerous, and often compiled by people who have no direct connection with budget decisions."

President George W. Bush's 2004 budget presented a new assessment tool, the Program Assessment Rating Tool (PART).[5] Unlike the mandate in GPRA, which focused on agencies, the emphasis of PART is on programs. PART examines four areas of assessment of an agency's programs through a set of approximately 30 questions to be completed by appropriate individuals within the respective agency program in conjunction with an OMB budget examiner.[6] These questions are focused on the following areas:

- *Program design and purpose*: whether the program's design and purpose are clear and defensible
- *Performance measurement, evaluation, and planning*: whether the program sets valid annual and long-term goals for its programs
- *Program management*: whether the program is effectively managed
- *Program results*: whether the program can report results with accuracy and consistency

Answers to each of the questions are self-reported using a 0 to 100 scale.[7] PART has been revised over time, and administrations have continued to emphasize program assessment through these revisions.[8]

Recently, however, there has been a renewed public interest in program evaluation. On October 7, 2009, Peter Orszag, then director of the Office of Management and Budget (OMB), sent a memorandum to the heads of executive departments and agencies on the subject of increased emphasis on program evaluations. Therein he wrote: "Rigorous, independent program evaluations can be a key resource in determining whether government programs are achieving their intended outcomes . . . . Evaluations can help policymakers and agency managers strengthen the design and operation of programs. Ultimately, evaluations can help the [Obama] Administration determine how to spend taxpayer dollars effectively and efficiently . . . ." (Orzag 2009).

As a second example of renewed public interest in program evaluation, the Science and Technology for America's Reinvestment: Measuring the Effect of Research on Innovation, Competitiveness and

Science (STAR METRICS) initiative was announced on June 1, 2010. It is a multi-agency effort led by the National Institutes of Health (NIH), the National Science Foundation (NSF), and the White House Office of Science and Technology Policy (OSTP) to help the federal government document the value of its investments in R&D.

3.3. Fiscal Accountability

Most of the legislation following GPRA emphasized fiscal accountability more than performance accountability. While it is not our intent to suggest that performance accountability is more or less important than fiscal accountability, for we believe that both aspects of public accountability are important, our emphasis herein is on performance accountability. Nevertheless, our discussion would not be complete in this chapter without references to the Government Management Reform Act of 1994 and the Federal Financial Management Improvement Act of 1996.

3.3.1. Government Management Reform Act of 1994

The Government Management Reform Act of 1994 builds on the Chief Financial Officers Act of 1990. Its purpose was to improve the management of the federal government though reforms to the management of federal human resources and financial management. Motivating the Act is the belief that federal agencies must streamline their operations and must reallocate their resources to better match a growing demand on their services. Government, like the private sector, must adopt modern management methods, utilize meaningful program performance measures, increase workforce incentives without sacrificing accountability, and strengthen the overall delivery of services.

3.3.2. Federal Financial Management Improvement Act of 1996

The Federal Financial Management Improvement Act of 1996 followed from the belief that federal agencies have not implemented federal accounting standards uniformly. Accordingly, this Act established a uniform accounting reporting system in the federal government.

The foregoing overview of what we call public accountability legislation shows government agencies are ostensibly becoming more accountable for their fiscal actions and their performance. These agencies are being required to a greater degree than ever before to account for their activities through a process of systematic measurement. For technology-based institutions in particular, internal difficulties are arising as organizations learn about that measurement process.

Compliance with the new accounting system is causing increased planning and impact assessment activity and is also stimulating greater attention to methodology. Perhaps there is no greater validation of this observation than the diversity of responses seen among public agencies, in general, and technology-based public institutions, in particular, as they struggle toward an understanding of the process of documenting and assessing their public accountability. Activities in recent years, apart from PART, have ranged from interagency discussion meetings to a reinvention of the assessment wheel, so to speak, in the National Science and Technology Council's (1996) report, "Assessing Fundamental Science."

Notes

1. This chapter draws from Link (1999), and Link and Scott (1998b) and the Link and Scott references therein.

2. Historically, according to Hufschmidt (1988), evaluation of public investment projects using benefit–cost concepts began in the United States during the Great Depression. Efforts were made to quantify the "New Deal" programs of President Franklin Roosevelt. Hufschmidt also carefully outlines landmarks in the evolution of benefit–cost analysis beginning with the issues addressed by the National Planning Board, established in 1934, through the 1980s.

3. Using the terminology that we proffered in Chapter 1, we would opt for the term *evaluation* rather than *assessment*, but as we also wrote in Chapter 1 the terms are often used interchangeably.

4. See, for example, Cozzens (1995).

5. In April 2002 a draft of PART was first tested on 67 programs, and the President's Management Council approved it for broader use in July 2002.

6. Stalebrink (2009) emphasizes that GPRA performance indicators are developed by the agency, whereas PART performance indicators are developed by OMB. In addition, GPRA's indicators are broader in scope, emphasizing various outcomes; PART's indicators are narrowly focused on effective outcomes, and we consider such indicators to be relevant to an assessment rather than an evaluation.

7. See OMB (2006).

8. In 2003, GAO undertook a study of PART, and, among other things, it concluded that PART is not well aligned in concept or in application with GPRA (GAO 2004). PART is not a substitute for GPRA's focus on comparisons across agencies.

4

Systematic Approaches to Program Evaluation and Evaluation Metrics

4.1. GPRA's Mandate

The Government Performance and Results Act (GPRA) was directionally, as opposed to methodologically, clear about the evaluation process.[1] It stipulated that public institutions/research programs must identify outputs and quantify the economic benefits of the outcomes associated with such outputs. In our opinion, agencies will attempt to quantify outcome benefits and then compare those quantified benefits to the public costs to achieve the benefits. Although these were GPRA's directions, the methodological hurdle that has been plaguing most public agencies has been how to quantify benefits. Furthermore, even with an acceptable quantification of benefits, will the confidence of the American people in public-sector research be strengthened by simply comparing benefits to costs?

We shall consider three different economics-based approaches—traditional, counterfactual, and spillovers methods—to program evaluation, and we shall also consider non-economics-based approaches.[2] We provide an overview of these approaches here; and then, in subsequent chapters, we illustrate each method with detailed case studies.

4.2. Traditional Economics-Based Evaluation Methods

Griliches (1958) and Mansfield et al. (1977) pioneered the application of fundamental economic insight to the development of estimates of private and social rates of return to investments in R&D. Streams of investment outlays through time—the costs—generate streams of economic surplus through time—the benefits. Once identified and measured, these streams of costs and benefits are used to calculate rates of return, benefit-to-cost ratios, or other related metrics.

To emphasize the impact that the Griliches/Mansfield model has had on the evaluation community, Figure 4.1 and Figure 4.2 illustrate graphically the cumulative non-self-citations to both the pioneering paper by Griliches in 1958 and that by Mansfield et al. in 1977, respectively. The citation information underlying each figure comes from the Social Science Citation Index, as accessed through the Web of Science. The citation information used in the figures relates to references to the evaluation methodology within each paper. Each time series begins in the year of the seminal publication and goes through the end of calendar year 2008.

Two points are worth noting in both figures. The first point is that the slope of each cumulative distribution is observably constant, which implies that the number of non-self-citations each year has remained fairly constant. Our interpretation of this observation is that

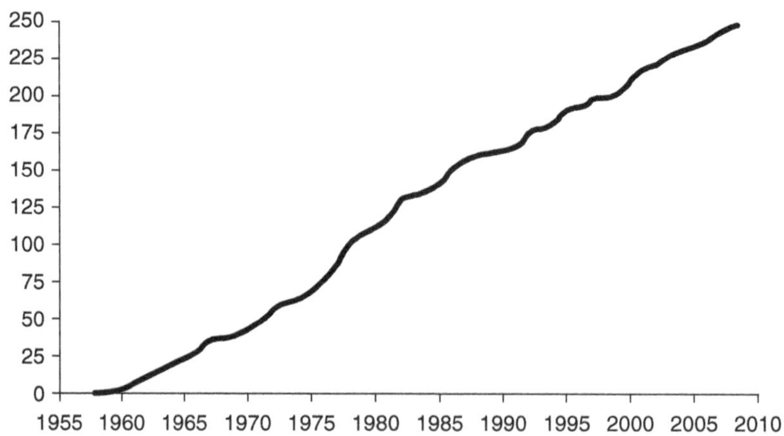

Figure 4.1 Cumulative distribution of non-self-citations to Griliches (1958).

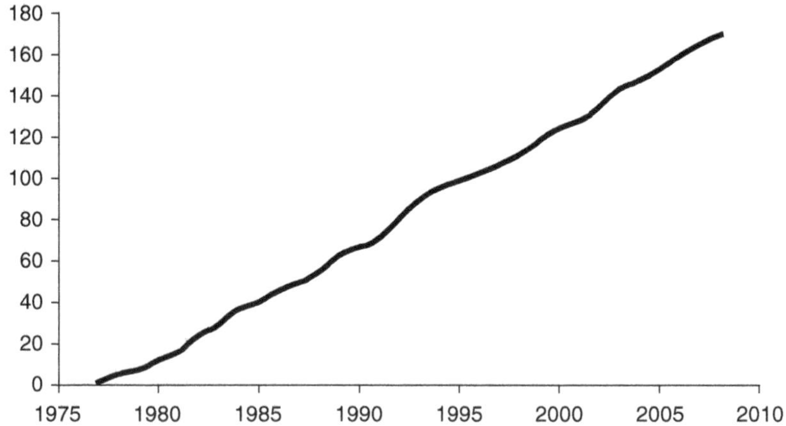

Figure 4.2 Cumulative distribution of non-self-citations to Mansfield et al. (1977).

the importance of these foundation papers has been long lived. The second point is that the total number of Web of Science non-self-citations to the evaluation methodology in each paper is modest by current standards in economics for recent papers that have helped to define a field of study—247 for Griliches (1958) and 170 for the later publication by Mansfield et al. (1977).[3] There are a number of possible interpretations of this observation, but one interpretation is that, in comparison with older scholars, many of the younger scholars who are engaged in the growing research and professional field of program evaluation are less aware of the hallmark literature.

In the Griliches/Mansfield approach's simplest model, the innovations evaluated are conceptualized as reducing the cost of producing a good sold in a competitive market at constant long-run unit cost as shown in Figure 4.3. For any period, there is a demand curve for the good and, in the simplest model, a horizontal supply curve. Innovation lowers the unit cost of production, shifting downward the horizontal supply curve and thereby, at the new lower equilibrium price, resulting in greater consumer surplus (the difference between the price consumers would have been willing to pay and the actual price they paid, summed over all purchases).[4]

Additionally, for market settings more complicated than the simplest model, the Griliches/Mansfield model accounts for producer surplus, measured as the difference between the price the producers receive per unit and the actual marginal cost, summed over the output sold,

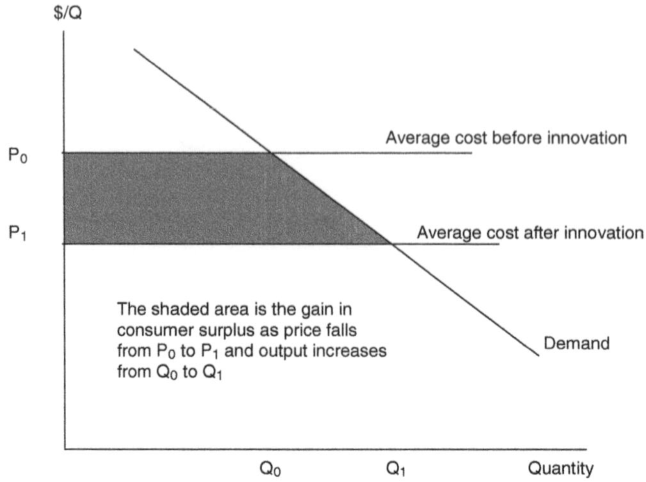

Figure 4.3 Gain in economic surplus from process innovation in a competitive market in the case of long-run constant unit costs.

minus any fixed costs. Social benefits are then the streams of new consumer and producer surpluses—economic value above and beyond the opportunity costs of the resources used to create value, while private benefits for a firm that invests in innovation are the portions of the streams of producer surplus appropriated by the investor. Not all of the appropriated producer surplus is necessarily new because the surplus gained by one producer might be cannibalized from the preinnovation surplus of another producer. Social and private costs will, in general, also be divergent. Figure 4.4 illustrates the economic surplus created each period for the simple case of an altogether new product (that does not cannibalize surplus from previously existing products) introduced by a monopolist.[5]

The Griliches/Mansfield model for calculating economic social rates of return adds the public and the private investments through time to determine social investment costs, and then the stream of new economic surplus generated from those investments is the benefit. Thus, the evaluation question that can be answered from such an evaluation analysis is: What is the social rate of return to the innovation, and how does that compare to the private rate of return?

We argue that this is not the most appropriate question to ask from a public accountability perspective. The fact that the social rate of return is greater than the private rate of return could validate the role of government in innovation if the private sector would not have undertaken

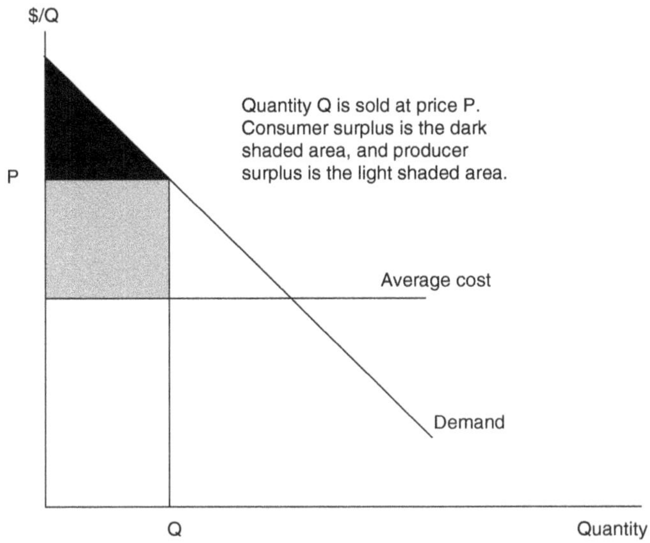

Figure 4.4 Consumer and producer surplus from a monopolist's innovative product.

the research; but it ignores, for example, consideration of the cost effectiveness of the public sector undertaking the research as opposed to the private sector.[6]

4.3. The Counterfactual Evaluation Method

A different question should be considered when publicly funded, publicly performed investments are evaluated. Holding constant the very stream of economic surplus that the Griliches/Mansfield model seeks to measure, and making no attempt to measure that stream, one should ask the counterfactual question: What would the private sector have had to invest to achieve those benefits in the absence of the public sector's investments? The answer to this question gives the benefits of the public's investments—namely, the costs avoided by the private sector.[7] With those benefits—obtained in practice through extensive interviews with administrators, federal research scientists, and those in the private sector who would have to duplicate the research in the absence of public performance—counterfactual rates of return and benefit-to-cost ratios can be calculated to answer the fundamental evaluation question: Are the public investments a more efficient way of generating the technology than private sector investments would have been? The answer to this question is more in line with the public accountability issues implicit in GPRA, and it is certainly more in line with the thinking of public sector stakeholders—or so we believe—who may doubt the appropriateness of government's having a role in the innovation process in the first place.

4.4. The Spillover Evaluation Method

There are important projects where economic performance can be improved with public funding of privately performed research. The question asked in the spillover method is one that facilitates an economic understanding of whether the public sector should be underwriting the private-sector firms' research, namely: What is the social rate of return from the program (including spillovers) compared to the private rate of return? Or: What proportion of the total profit stream generated by the private firm's R&D and innovation does the private firm expect to capture; and hence, what proportion is not appropriated but is instead captured by other firms that imitate the innovation or use knowledge generated by the R&D to produce competing products for the social good? The part of the stream of expected profits captured by the innovator is its private return, while the entire stream is the lower bound on the social rate of return. In essence, this method weighs the private return, estimated through extensive interviews with firms receiving public support regarding their expectations of future patterns of events and future abilities to appropriate R&D-based knowledge, against private investments. The social rate of return then weighs the social returns against the social investments.

The application of the spillover model to the evaluation of publicly funded, privately performed research is appropriate because the output of the research is only partially appropriable by the private firm with the rest spilling over to society. The extent of the spillover of such knowledge with public good characteristics determines whether the public sector should fund the research.

4.5. Non–Economics-Based Evaluations

There will sometimes be cases for which it will not be possible to collect economic data about benefits and costs needed to develop the economics-based traditional, counterfactual, and spillovers evaluation methods. In such cases, non-economics-based evaluation approaches provide bibliometric documentation of the outputs from a research project—counts of outputs such as the number of publications, citations, patents, and presentations produced by the project[8]—as well as documentation of the scope and pace of the project. Such evaluations typically cover a portfolio of research projects and begin with a survey of the projects' principal investigators and available information about the projects.

We shall emphasize three keys to the success of such evaluations, although these points are applicable to other approaches as well as non-economics-based models. The first key is careful development of the counterfactual to which the non-economics-based model's metrics are to be compared. A public-supported research project's outputs and the pace of its progress can be compared with what would have occurred in the absence of the public support. This point applies also to the implementation of the counterfactual evaluation method discussed earlier. It is important to know what technology would have prevailed in the absence of the technology-based program being evaluated. In other words, the counterfactual situation must reflect the frontier state of the art that would have prevailed.

The second factor relates to attribution. In the case of output counts (e.g., publications), care must be taken to ensure that the counted outputs are fully attributable to the program being evaluated. Often, intellectual property is a result of cumulative research, some of which could have been funded from sources other than those being studied. Attribution is equally important in the implementation of the counterfactual evaluation method; specifically, it could be that measured benefits are not fully the result of the investments made by the program being evaluated but rather only partially attributable to its investments. There may have been other public investments; for example, one publicly performed research program may obtain some of its resources from another public program. Typically R&D-based technology programs are public–private partnerships, and the private sector will typically incur investment costs that together with the public's investments determine the stream of benefits from the program. In such cases, when evaluation metrics are developed, the benefits created by the program (and the extra-program efforts that supported its work) should be weighed against all of the social costs incurred to create those benefits. Or, when a research program builds on the results of earlier programs, such as when a new standard reference material is developed by a federal laboratory, the evaluation of the benefits should develop expert opinion about the incremental contributions of the new infrastructure technology—what it added to the previous state of the art—weighing just those incremental benefits against the social cost of the program.

A third critical factor is careful control for the selection into the studied sample by response to the survey. Nonmarket counterfactual evaluation with control for response bias allows evaluation of project and program performance when market information is unavailable, and it provides useful additional information for assessments even when market valuations are available.

Although not grounded in economic (or other) theory, bibliometric counts do have the benefits of being a flexible characterization of scholarly activity at various levels of disaggregation. They can be used to compare the productivity of researchers both over time and to each other; they can be used similarly to describe the productivity of programs, institutions, as well as countries. However, selectivity problems remain about what is being counted and/or where it is being cited.

4.6. Evaluation Metrics

Broadly viewed, there are two types of evaluation metrics: *(1)* non–economics-based metrics, including bibliographic measures, and *(2)* metrics grounded in economic theory. The non–economics-model-based measures require no elaboration here, and we illustrate such measures in Chapter 6. Here we explain the economics-based metrics commonly used in the evaluation literature.[9]

For an evaluation of a public R&D-based program, two time series of data are needed. One time series is for the constant-dollar costs associated with the program, public costs, and private costs. The other time series is for the constant-dollar benefits to those whose R&D is being leveraged. Several metrics are common in evaluations, especially when the traditional or counterfactual evaluation method is used.

4.6.1. Internal Rate of Return

The internal rate of return (*IRR*)—a real rate of return in the context of constant-dollar cash flows—is the value of the discount rate, i, that equates the net present value (*NPV*) of the stream of net benefits associated with a research project to zero. The time series runs from the beginning of the research project, $t = 0$, through a terminal point, $t = n$.

Mathematically,

$$NPV = [(B_0 - C_0)/(1 + i)^0] + \ldots + [(B_n - C_n)/(1 + i)^n] = 0 \quad (4.1)$$

where $(B_t - C_t)$ represents the net benefits associated with the project in year t, and n represents the number of time periods—years in the case studies below—being considered in the evaluation.

For unique solutions for i, from equation (4.1), the *IRR* can be compared to a value, r, that represents the opportunity cost of funds invested by the technology-based public institution. Thus, if the opportunity cost of funds is less than the internal rate of return, the project was worthwhile from an ex post social perspective.[10]

4.6.2. Benefit-to-Cost Ratio

The ratio of benefits-to-costs (B/C) is the ratio of the present value of all measured benefits to the present value of all measured costs. Both benefits and costs are referenced to the initial time period, $t = 0$, when the project began as:

$$B/C = [\Sigma_{t=0 \text{ to } t=n} B_t/(1 + r)^t]/[\Sigma_{t=0 \text{ to } t=n} C_t/(1 + r)^t] \quad (4.2)$$

A benefit-to-cost ratio of 1 is said to indicate a project that breaks even. Any project with $B/C > 1$ is a relatively successful project as defined in terms of benefits exceeding costs.

Fundamental to implementing the ratio of benefits-to-costs is a value for the discount rate, r. While the discount rate representing the opportunity cost for public funds could differ across a portfolio of public investments, the calculated metrics in the case studies in this book follow the guidelines set forth by the Office of Management and Budget (OMB, 1992) in Circular A-94, which states that: "Constant-dollar benefit-cost analyses of proposed investments and regulations should report net present value and other outcomes determined using a real discount rate of 7 percent."[11]

4.6.3. Net Present Value

The information developed to determine the benefit-to-cost ratio can be used to determine net present value (*NPV*) as:

$$NPV_{\text{initial year}} = B - C \quad (4.3)$$

where, as in the calculation of B/C, B refers to the present value of all measured benefits, C refers to the present value of all measured costs, and present value refers to the initial year or time period in which the project began, $t = 0$ in terms of the B/C formula in equation (4.2). Note that *NPV* allows, in principle, one means of ranking several projects ex post, providing investment sizes are similar.

4.7. Conclusions

The metrics discussed in section 4.6 are the evaluation tools most commonly used when summarizing the findings from evaluations based on economics-based methods. The benefit-to-cost ratio and net present value are obviously related metrics, but the use of these metrics requires additional discussion.

Benefit–cost analysis is a decision-making tool that can be, and has been, applied to many different situations related to public sector activity. Prospective analyses are generally of two types: one is related to the decision to invest or not to invest in a particular project, and the other is related to the choice between two defined possible projects. Retrospective analyses are generally involved with determining whether the choice of a prior project was economically sound. In either application, there are those who are critical of benefit–cost analysis because of its shortcomings, but the critics do not offer unflawed or complete alternatives. A benefit–cost analysis sets out views of the benefits and costs that can be discerned, debated, and modified if discussion determines appropriate changes. Benefit–cost analysis thus provides a key tool for sound policy.

As Frank (2000) and Sen (2000), among others, have pointed out, benefit–cost analysis often involves a comparison of disparate and incommensurate categories of benefits, sometimes artificially denominated into dollars. And benefit–cost analysis often ignores distribution effects, meaning that there is no attempt to deal with the fact that the benefits and the costs are not realized equally across different groups, for example, by income level or demographics.[12]

Notes

1. Link (1996a, 1996b) raised this issue early in the discussions about GPRA.
2. This section draws from Link and Scott (1998b, 2000, 2005a).
3. Google Scholar reports about 1,300 citations to Griliches (1958) and nearly 400 to Mansfield et al. (1977). Google Scholar searches more broadly (e.g., books

and book chapters) than the Web of Science, and it does not distinguish between citations to methodology as opposed to citations to findings. It is not the specific number of citations that is key, in our opinion; rather, it is the constancy with which the methodology in these classical articles has been referenced over time.

4. The diagrammatic treatment of consumer surplus shows an approximation of the increase in economic welfare. See Willig (1976) for an explanation of why we expect the approximation to be a good one.

5. See Scherer (1979) for discussion, estimation of the surplus gained, diagrams of the more complex case where previously existing surplus is cannibalized by the new product, and additional references to the related literature.

6. That said, public agencies might request that the Griliches/Mansfield model be applied, as illustrated in the case study in Chapter 7. With the fundamental publicly performed R&D evaluation question—that is, was public performance of the R&D efficient as compared with having the private sector do it?—answered with the counterfactual method to be described in the text that follows, using the Griliches/Mansfield evaluation method as well complements the evaluation. It answers the question of whether—assuming the private sector would not have undertaken the project or simply ignoring the possibility that it would have—the publicly performed R&D investment generates a stream of economic surplus with a value greater than its costs.

7. Stated more completely, the benefits from the public sector, rather than the private sector, performing the investments should be measured. The counterfactual method measures as benefits the private sector's costs avoided through the public's investments plus the benefits from the public sector's investments that industry would be unable or unwilling to duplicate. In the extreme case where industry would not have made the investments at all, there are no private-sector costs avoided, but because the private-sector performance shortfall is complete, all of the traditional Griliches/Mansfield stream of returns to the R&D investments are valued as benefits. In that special case, the counterfactual approach is identical to the Griliches/Mansfield approach except that it has the advantage of having pointed out that government could do the work more efficiently—in this special case because industry would not do it at all.

8. Narin (1976) pointed out that publication- and citation-counting techniques have been used in the assessment of scientific activity for decades, tracing at least to Cole and Eales (1917). Narin et al. (1994) noted that the first use of publication, patent, and citation counts as performance indicators was by the National Science Board (1972).

9. See Ruegg and Feller (2003) for an important applied interpretation of these metrics.

10. For the archetypal illustrative case to explain the concept of internal rate of return, net cash flows of benefits minus the costs begin as negative and then become positive, with just a single reversal in the sign for the series of flows. There is then at most a single real solution for the internal rate of return. Of course, for actual investment projects there can be multiple reversals in the signs for the net cash flows, and then using the internal rate of return concept for a benefit–cost analysis requires some additional work with the time series of benefits and costs. For an actual example with an explanation of how to treat the issue, see Link and Scott (1998b, p. 46).

11. In Chapter 2, we have noted and discussed the OMB's choice, as explained in OMB (2003), of 7% as the social hurdle rate. Circular A-94 (OMB

1992) was originally issued under the authority of the Budget and Accounting Act of 1921 (discussed in Chapter 3).

12. Scott (2009b) provides a detailed discussion and development of the reasons that the market-centered valuations for costs and benefits that are customarily used in benefit–cost analyses can be appropriate, even though the market-centered valuations do not necessarily reflect the preferences of those lacking the income to buy the goods and services provided by the evaluated R&D investments.

5

Introduction to the Case Studies

5.1. Four Case Studies and Four Evaluation Methods

Four case studies are discussed in Chapters 6–9, each using one of the four evaluation methods discussed in Chapter 4. We are privileged to have previously conducted these case studies. Thus, each is dated in time but not in methodological applicability. In each case study we provide detailed institutional background, including a discussion of the relevant technology, and an interpretation of the evaluation metrics. Each of the case studies is relevant to research activities at the National Institute of Standards and Technology (NIST), and each is relevant to the conduct and provision of infrastructure technology.

The first case study provides an evaluation of the intramural program within NIST that allocated some of the Advanced Technology Program's (ATP's) budget for research conducted within NIST's laboratories in support of the technology initiatives of ATP. Non-economics-based evaluation tools, including bibliometric evaluations, were used to evaluate the intramural program.

The second case study is about intramural research that took place at NIST to develop an improved standard reference material (SRM) for the measurement of the wavelength of light in an optical fiber network. Because standard reference materials are publicly funded and the underlying research is publicly performed, the counterfactual evaluation method is appropriate. However, the traditional evaluation method was used at the request of the sponsoring agency to evaluate the net social benefits of this SRM. As explained in Chapter 4, the information developed with the traditional evaluation method complements the information developed using the counterfactual method.

The third case study is for the Malcolm Baldrige National Quality Award Program, also within NIST. The counterfactual evaluation method was used to evaluate the net social benefits from this program because it is a publicly funded program and the evaluation criteria are publicly developed.

The fourth case study is about a focused program sponsored by ATP on Technologies for the Integration of Manufacturing Applications (TIMA). The spillover evaluation method was used to evaluate the net social benefits from this focused program. TIMA technologies were publicly funded, but the research was privately performed.

The four case studies are summarized in Table 5.1 in terms of technology and evaluation method.

5.2. Background Information on NIST

NIST is a federal laboratory.[1] Its current activities are perhaps best understood in terms of its mission of providing a public good: voluntary

Table 5.1 Summary of the Case Studies

Chapter	Technology	Funding Program	Evaluation Method
6	Intramural research program	ATP	Bibliometric analysis of publication and patent counts
7	Wavelength references for optical fiber communications	ATP	Traditional Griliches/ Mansfield approach
8	Malcolm Baldrige National Quality Award	NIST	Application of the counterfactual method to publicly funded, publicly performed research
9	Technologies for the integration of manufacturing applications	ATP	Application of the spillover method to publicly funded, privately performed research

ATP, Advanced Technology Program; NIST, National Institute of Standards and Technology.

standards for industry and commerce. A standard (and in particular, given NIST's mission, a standard for industry and commerce) is a set of specifications to which all elements of products, processes, formats, or procedures under its jurisdiction must conform.

The concept of the government's involvement in standards traces to the Articles of Confederation signed on July 9, 1778. Article 9, §4 stated: "The United States, in Congress assembled, shall also have the sole and exclusive right and power of regulating the alloy and value of coin struck by their own authority, or by that of the respective States; fixing the standard of weights and measures throughout the United States . . . ." This responsibility was reiterated in Article 1, §8 of the Constitution of the United States: "The Congress shall have power . . . To coin money, regulate the value thereof, and of foreign coin, and fix the standard of weights and measures. . . ."

On July 20, 1866, Congress and President Andrew Johnson authorized the use of the metric system in the United States. This was formalized in the Act of July 28, 1866: An Act to Authorize the Use of the Metric System of Weights and Measures.[2] Within a year, a standardized measurement plan was adopted based on terrestrial arcs, and the term *mètre* (meter), from the Greek *metron* meaning to measure, was assigned by the Academy of Sciences (Link 1996b).

Because of the growing use of the metric system in scientific work rather than commercial activity, the French government held an international conference in 1872, which included the participation of the United States, to settle on procedures for the preparation of prototype metric standards. Then, on May 20, 1875, the United States participated in the Convention of the Meter in Paris and was one of the 18 signatory nations to the Treaty of the Meter. In a Joint Resolution before Congress on March 3, 1881, it was resolved that: "The Secretary of the Treasury . . . is hereby directed to cause a complete set of all the weights and measures adopted as standards to be delivered to the governor of each State in the Union. . . ." Then, the Act of July 11, 1890 gave authority to the Office of Construction of Standard Weights and Measures (or Office of Standard Weights and Measures), which had been established in 1836 within the Treasury's Coast and Geodetic Survey.

Following from a long history of our nation's leaders calling for uniformity in science, traceable at least to the several formal proposals for a Department of Science in the early 1880s, and coupled with the growing inability of the Office of Weights and Measures to handle the explosion of arbitrary standards in all aspects of federal and state activity, it was inevitable that a standards laboratory would need to be established. The political force for this laboratory came in 1900 through Lyman Gage, then Secretary of the Treasury under President William McKinley.

Gage's original plan was for the Office of Standard Weights and Measures to be recognized as a separate agency called the National Standardizing Bureau. This Bureau would maintain custody of standards, compare standards, construct standards and test standards, and resolve problems in connection with standards. Although Congress at that time wrestled with the level of funding for such a laboratory, the importance of the laboratory was not debated.

Finally, the Act of March 3, 1901, also known as the Organic Act, established the National Bureau of Standards within the Department of the Treasury, where the Office of Standard Weights and Measures was administratively located:

> *Be it enacted by the Senate and House of Representatives of the United States of America in Congress assembled,* That the Office of Standard Weights and Measures shall hereafter be known as the National Bureau of Standards . . . That the functions of the bureau shall consist in the custody of the standards; the comparison of the standards used in scientific investigations, engineering, manufacturing, commerce, and educational institutions with the standards adopted or recognized by the Government; the construction, when necessary, of standards, their multiples and subdivisions; the testing and calibration of standard measuring apparatus; the solution of problems which arise in connection with standards; the determination of physical constants and the properties of materials, when such data are of great importance to scientific or manufacturing interests and are not to be obtained of sufficient accuracy elsewhere.

The Act of February 14, 1903 established the Department of Commerce and Labor, and in that Act it was stated that the National Bureau of Standards be moved from the Department of the Treasury to the Department of Commerce and Labor. Then, in 1913, when the Department of Labor was established as a separate entity, the Bureau was formally housed in the Department of Commerce.

In the post–World War I years, the Bureau's research focused on assisting in the growth of industry. Research was conducted on ways to increase the operating efficiency of automobile and aircraft engines, electrical batteries, and gas appliances. Work was also begun on improving methods for measuring electrical losses in response to public utility needs. This latter research was not independent of international efforts to establish electrical standards similar to those established over 50 years earlier for weights and measures.

After World War II, significant attention and resources were given to the activities of the Bureau. In particular, the Act of July 21, 1950 established standards for electrical and photometric measurements. Then, as a part of the Act of June 20, 1956, the Bureau moved from Washington, D.C., to Gaithersburg, Maryland. The responsibilities listed in the Act of July 21, 1950, and many others, were transferred to NIST when the National Bureau of Standards was renamed under the guidelines of the Omnibus Trade and Competitiveness Act of 1988:

> The National Institute of Standards and Technology [shall] enhance the competitiveness of American industry while maintaining its traditional function as lead national laboratory for providing the measurement, calibrations, and quality assurance techniques which underpin United States commerce, technological progress, improved product reliability and manufacturing processes, and public safety . . . [and it shall] advance, through cooperative efforts among industries, universities, and government laboratories, promising research and development projects, which can be optimized by the private sector for commercial and industrial applications . . . [More specifically, NIST is to] prepare, certify, and sell standard reference materials for use in ensuring the accuracy of chemical analyses and measurements of physical and other properties of materials. . . .

NIST's mission is to promote U.S. economic growth by working with industry to develop and apply technology, measurements, and standards. It provides technical leadership for vital components of the nation's technology infrastructure needed by U.S. industry to continually improve its products and services. NIST's organizational structure is laboratory based; currently, there are seven laboratories and two research centers.[3] These laboratories and centers provide technical leadership for vital components of the nation's technology infrastructure needed by U.S. industry to continually improve its products and services.

5.3. Background Information on Infrastructure Technology

In Chapter 2 we briefly mentioned infrastructure technology as an activity of a public research institution, with the example being the promulgation and adoption of standards. Here we discuss infrastructure technology in greater detail to provide background for understanding the research being evaluated in the case studies that follow in subsequent chapters. Infrastructure technologies, or infratechnologies, can de defined as:

> technologies that *support* R&D, production, and marketing in industries. . . . Infratechnologies include evaluated scientific data used in the conduct of R&D; measurement and test methods used in research, production control, and acceptance testing for market transactions; and various technical procedures such as those used in the calibration of equipment. Infratechnologies are not generally embodied in an industry's product technology in the same way as the generic technology (the knowledge that is organized into the conceptual form of an eventual application).
>
> Instead, infratechnologies facilitate development of the generic technology by, for example, providing highly precise measurements and creating organized and evaluated scientific and engineering data necessary for understanding, characterizing, and interpreting relevant research findings. Typically, they tie, at root, to the fundamental units of measurement. They also provide the measurement and testing concepts and techniques that enable higher quality and greater reliability at lower cost in production. Finally, infratechnologies provide buyers and sellers mutually acceptable, low-cost methods of assuring that specified performance levels are met when technologically sophisticated products enter the marketplace. (Link and Tassey 1987, pp. 18–19)

Hence, infrastructure technologies are tools and procedures—products and services—that embody practical applications of knowledge and that are used by science and industry to facilitate the capabilities and methods of implementing and using further practical applications of knowledge (i.e., other technologies). Infrastructure technologies are therefore, as Tassey (2008) emphasizes, the technological foundation resources necessary to support the full range of scientific and commercial activities, including research and development, production, and successful marketing of final products.[4]

Examples of infrastructure technology include standard reference materials, standard reference data, calibrations services, traceability of measurements, and supporting standards-related programs, knowledge, and services. NIST is the primary public provider of infrastructure technology in the United States. Table 5.2 lists some, but by no means all, of the interrelated and mutually reinforcing infrastructure technology products and services that NIST provides, along with a brief description of each.

Infrastructure technology is crucial for R&D investment and productivity. As Tassey (2008) observes:

> Infrastructure in the form of a wide variety of infratechnologies and associated standards is essential to conduct R&D, control production processes for quality and yield, and finally to execute market transactions for complex products.
>
> Examples of measurement infratechnologies are measurement and test methods, process and quality control techniques, evaluated scientific and engineering data, and the technical basis for product interfaces.
>
> Imagine trying to conduct R&D without critically evaluated data on the properties of advanced materials, or accurate and consistent measurement techniques, controlling production processes to attain high yields and quality levels without sensors that instantly and accurately measure the performance attributes of a product as it is being manufactured. Because these needs are essential, most high-tech industries are measurement-intensive. (pp. 1–2)

Successful marketing of products requires infratechnologies as well. As Tassey (2008) observes:

> When interoperability standards and the supporting infratechnologies are inadequate, companies must expend considerable resources to avoid compatibility problems or mitigate problems after they occur. Moreover, such activities can result in delays with respect to product/service delivery, which adds additional costs.

Table 5.2 Selected NIST Infrastructure-Technology Products and Services

Product or Service	Brief Description
Calibrations	NIST calibration services help makers and users of precision instruments to achieve the highest possible levels of measurement quality and productivity. These services directly link a customer's instrument or transfer standard to national and international standards.
Databases	NIST customers can purchase a wide variety of highly accurate scientific and technical databases through the NIST Standard Reference Data Program. A number of popular NIST data collections, such as the Chemistry WebBook and Physical Reference Data, are available for free on the World Wide Web.
Laboratory accreditation	Public and private laboratories can demonstrate their competence to perform tests and calibrations in a specific field by earning accreditation in the National Voluntary Laboratory Accreditation Program.
Measurement and standards research	Researchers in NIST Laboratories, located in Gaithersburg, Maryland, and Boulder, Colorado, conduct research in physical and engineering sciences. The labs respond to industry needs for measurement methods, tools, data, and technology. NIST researchers collaborate with colleagues in industry, academic institutions, and other government agencies.
Software	NIST provides many freely available software programs, including, for example, programs for economic analysis of life-cycle building costs and choices among new technologies, for modeling the structure and properties of cement-based materials, for conformance testing of software to measure whether a product implements an information technology specification, for smoke plume modeling to predict the downwind distribution of smoke particulate and combustion products from large outdoor fires, and to allow personal computers to set their clocks to the NIST atomic clock.
Standard reference materials (SRMs)	A standard reference material is a NIST-produced measurement tool designed to ensure the accuracy of measurements in industry, manufacturing, medicine, environmental monitoring, and research. NIST produces and sells industry, academia, government, and other users over 1,300 reference materials of the highest quality and metrological value.

Table 5.2 *Continued*

Standards	The Standards Services Division conducts standards-related programs and provides knowledge and services with the goals of ensuring recognition and use of U.S. standards domestically and in the global marketplace; promoting worldwide acceptance of U.S. test and calibration data to facilitate the marketing of U.S. products; and providing assistance to industry, trade associations, and exporters through its operations, training, and information programs.
Traceability	NIST is responsible for developing, maintaining, and disseminating national standards for the basic measurement quantities, and for many derived measurement quantities. NIST is also responsible for assessing the measurement uncertainties associated with the values assigned to these measurement standards. As such, the concept of measurement traceability is central to NIST's mission. NIST helps other organizations establish traceability of their measurement results to standards developed and maintained by NIST.
Weights and measures	NIST's Office of Weights and Measures works to assure a uniform system of weights and measures in the United States. Accuracy in weights and measures gives consumers confidence in the marketplace and assures fair competition among businesses.

NIST, National Institute of Standards and Technology.
Source: NIST's Web site (http://www.nist.gov).

> Access to targeted markets . . . increasingly depends on providing not only an innovative product but also including a technically sophisticated interface to other components of the product system. Engineering complex, non-standard interfaces, which often vary significantly among customers, adds both excessive costs and time delays. . . .
>
> Even the act of buying a high-tech product requires a technically sophisticated infrastructure. For example, how would a communications services company determine if the thousands of kilometers of ultra-thin optical fiber it needs to buy meet its specifications for such performance attributes as band width, signal loss rates, core diameter consistency, insulation properties, etc. unless technologically advanced acceptance testing standards are available? Without such infrastructure, the high-tech sector of the economy could not grow and prosper.

> Market transaction costs are substantially higher without adequate supporting infrastructure and time delays can eliminate the "first mover" advantage of innovating firms. (pp. 4–5)

Why is infrastructure technology so important for R&D investment? The answer lies in the economics of investment in R&D, the profitability of which depends first on the technical success of the R&D program and then on the ultimate success of an R&D investment's output when it is embodied through production in products and then when the ensuing final products and services are marketed. Key are *(1)* barriers to technology—circumstances where private rates of return from technology investments fall short of private hurdle rates (i.e., private required rates of return), yet social rates of return exceed the social hurdle rate, and the private shortfall in returns causes underinvestment in infrastructure technology and R&D—and *(2)* the stimulating effect of publicly provided infrastructure technology on R&D investment and productivity.

The barriers to technology result in a market failure of insufficient private investment in infrastructure technology. Then, in the absence of sufficient infrastructure technology, there is, from society's perspective, underinvestment in R&D more generally, resulting in a shortfall of socially valuable productivity gains from their potential.

Economic theory and evidence, taken together, show a need for public provision and support of infrastructure technology. Its benefits are available to many firms, even if they did not contribute to the investments in the technology. In the theoretical case where the return on private investment in infrastructure technology would be sufficient to justify it, even though the private investors did not appropriate all of the social benefits resulting from their investment, it is possible that private firms would invest in sufficient infrastructure technology. However, we do not expect individual firms will invest in infrastructure technology to the socially optimal extent for theoretical reasons that are supported by the evidence.

We expect the private sector alone would underinvest in infrastructure technology for theoretical reasons that stem from the observation, as Tassey (2008) emphasizes, that the technology is a quasi-public good. Infrastructure technology has some of the attributes of a public good. Like a public good, infrastructure technology is nonrivalrous in its use (i.e., one firm can use the technology and it is still there for others to use). Furthermore, although not as completely as for a pure public good, to some extent infrastructure technology is nonexcludable (i.e., some of the benefits are available to firms even if they do not pay for the use of the technology). Consequently, it would be difficult for private firms profitably to provide and market infrastructure technology. They would incur the investment costs, but their investment would create positive externalities—benefits of the investment would not be appropriated by the investors but instead would spill over to others who had not contributed to the investment yet would "free ride" and benefit from the investments of others. Moreover, infrastructure technology entails associated standards, and it is more valuable the wider the acceptance and use of the standards. Thus, there are network externalities—positive externalities in the use of the infrastructure technologies as well; hence, even if a user had to pay for the use, it would not be willing to pay the full value of its use because some of the value accrues to other users.

Theoretically it is possible that private investment would be sufficient despite the positive externalities, because although not complete, the appropriation of returns could be sufficient to allow an acceptable return on investment. However, a large body of evidence shows that public investments in infrastructure technology provide socially valuable investments that would not have been undertaken by the private sector, or would not have been undertaken to the same extent, or would have had higher investment costs and been less effective. Link and Scott (1998b, 2005a) provide many case studies of the socially valuable economic impact of public investments in infrastructure technologies. Tassey (2008) reviews many other case studies that have reached the same conclusion that public investment in infrastructure technology is needed to offset the private sector's underinvestment. In all, there is much evidence that supports the theoretical expectation that public investment is necessary to avoid underinvestment, and also inefficient investment, in infrastructure technology.

The barriers to technology that cause the private sector to underinvest in infrastructure technology are the same as were outlined in Table 2.1 in Chapter 2, with respect to factors creating barriers to innovation and technology. Public investment in infrastructure technology is crucial for R&D investment because *(1)* barriers to technology would otherwise result in underinvestment in infrastructure technology and *(2)* the publicly provided infrastructure technology has a stimulating effect on R&D investment and productivity.

Now, how is it that the public provision of infrastructure technology works to increase R&D investment and productivity? Infrastructure technology increases R&D and productivity *(1)* by increasing the value of the results, relative to the anticipated state of the art, of R&D investment and *(2)* by shifting the probability distribution for the quality of the technical results, relative to the anticipated state of the art, for any given R&D investment (e.g., the speed of an innovative computer relative to the anticipated state of the art). Value increases as a function of the relative success of R&D's technical results, and infrastructure technology shifts upward that value function. Thus, publicly provided infrastructure technology will have what Scott (2009a) identifies as value-shifting and probability-shifting effects on a firm's R&D investments. Because of its positive effects on production processes and its positive effects on commercial value, infratechnology increases the value of any R&D outcome relative to the anticipated state of the art and thus increases the marginal value of doing more R&D to shift the probability distribution of relative technical success rightward over better relative outcomes. Because of its positive effects on the R&D effectiveness of all firms' R&D investments, infratechnology also increases the state of the art expected from best practice technology. Hence, at the level of R&D that the firm would have chosen without the infratechnologies, those technologies may decrease the relative success (i.e., the quality relative to the anticipated state of the art) of any given technical outcome of a firm's R&D, and thus increase the firm's marginal value of doing more R&D to improve relative technical outcomes.[5] Importantly, the R&D-increasing effects that are caused by a value-shifting and probability-shifting variable—here infrastructure technology—obtain for both oligopolistic and competitively structured R&D environments.[6] The generality of the R&D-increasing effect of infrastructure technology is important because the types of competition in innovation markets vary.

NIST's infrastructure technology products and services provide the measurement-intensive infrastructure technology necessary to support the research and development activities in industry, academia, and government laboratories. The measurement infratechnologies and their associated standards will increase the value of a firm's technical R&D results because there will be less uncertainty about what those results are, production processes embodying the results will be more carefully controlled for quality and yield, and buyers of products embodying the R&D results will be able to ascertain the characteristics of the products and know how the products will perform. Such value-increasing effects increase the marginal value of R&D investment, and the firm will invest more. Moreover, the availability of publicly provided infrastructure technology will improve the pace and quality of research and development efforts, improving the expected state of the art emerging from the R&D investment competition among all investing firms. Consequently, the relative technical quality is less for any given R&D outcome for the firm. If the distribution for relative technical success has shifted leftward at the level of R&D that would be chosen without the infratechnology, then the firm has an incentive to invest more in R&D to improve the relative quality of its outcome.

Thus, economic analysis explains why publicly supported provision of infrastructure technology is necessary to avoid market failure of underinvestment in both the infrastructure technologies and R&D. Public investment in infrastructure technology is expected to increase socially valuable R&D; as Link and Siegel (2003) document, the increase in R&D will ultimately increase productivity growth. Figure 5.1 illustrates the link from public investment in infrastructure technology to increased R&D and then to productivity growth.

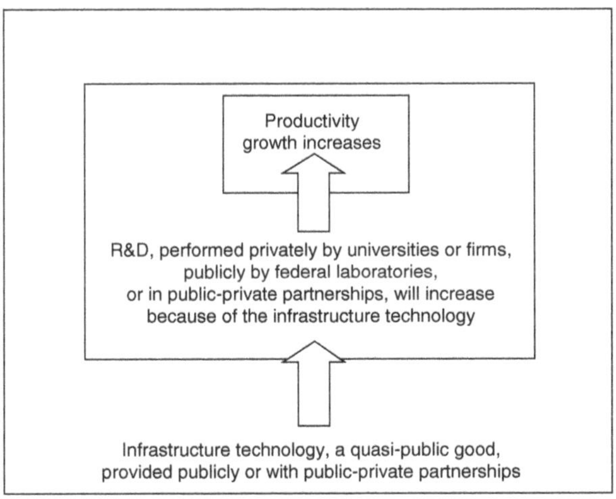

Figure 5.1 Infrastructure technology increases R&D and productivity growth.

Notes

1. This chapter draws on Link and Scott (1998b) and the references therein.

2. As background to this Act, the origins of the metric system can be traced to the research of Gabriel Mouton, a French vicar, in the late 1600s. His standard unit was based on the length of an arc of 1 minute of a great circle of the earth. Given the controversy over this measurement, the National Assembly of France decreed on May 8, 1790 that the French Academy of Sciences along with the Royal Society of London deduce an invariable standard for all the measures and all the weights.

3. Chapter 6 provides a more complete description of the NIST research laboratories.

4. "Infratechnology includes certain methods and basic data used throughout the process of technology-based competition: conducting R&D, product engineering, production, and marketing . . . . [I]ndividual firms typically use the pool of infratechnology (materials characteristics, measurements used for quality control processes, and various test methods) to create advantages in the marketplace" (Link and Tassey 1987, p. 84). Link and Tassey (1987) provide a full discussion of the distinctions and relations among the science base, generic technology, infratechnology, and proprietary technology.

5. Infrastructure technology increases the R&D effectiveness of all firms' R&D and therefore increases the state of the art anticipated from best practice. Thus, the infrastructure technology decreases the relative quality—that is, the relative technical success—of any given technical outcome from R&D. If the firm invested the same amount in R&D as it would in the absence of the public infrastructure and that resulted in a distribution over relative quality of technical outcomes that has shifted leftward because of the better anticipated state of the art, that leftward shift in the distribution increases the firm's marginal value of doing more R&D. More R&D is now valuable because without it the firm's relative position is less good. If instead the firm's relative technical success at the optimal level of R&D absent the infrastructure technology improves given that technology (i.e., if the probability distribution for the firm's relative technical success shifts rightward), then its optimal R&D investment will decrease. Of course, R&D is still more efficient, and there is greater productivity than without the infrastructure technology, but R&D is so much more effective for the firm relative to its competitors that it can do less and achieve the greatest expected profit. This is a partial effect, so even in the case that the firm's position relative to its competitors improves, the positive value-shifting effect could offset the probability-shifting effect here. With a leftward shift for the probability distribution given the increased effectiveness of all firms' R&D, both effects work together to increase a firm's optimal R&D investment. The foregoing discussion assumes that in the absence of infrastructure technology, R&D is nonetheless profitable. If that were not the case and the addition of appropriate infrastructure technology enabled profitable R&D, then of course R&D would increase with the provision of the infratechnology.

6. The effects of value-shifting and probability-shifting variables in the alternative types of R&D competition are established in Results 1, 2, 3, and 4 of Scott (2009a).

6

The Advanced Technology Program's Intramural Research Awards Program

6.1. Introduction

The Advanced Technology Program (ATP) was established within the National Institute of Standards and Technology (NIST) through the Omnibus Trade and Competitiveness Act of 1988 and modified by the American Technology Preeminence Act of 1991.[1] The goals of the ATP, as stated in its enabling legislation, were to assist U.S. businesses in creating and applying the generic technology and research results necessary to "(1) commercialize significant new scientific discoveries and technologies rapidly [and] (2) refine manufacturing technologies."

These same goals were restated in the *Federal Register* on July 24, 1990: "The ATP . . . will assist U.S. businesses to improve their competitive position and promote U.S. economic growth by accelerating the development of a variety of pre-competitive generic technologies by means of grants and cooperative agreements."

ATP received its first appropriation from Congress in fiscal year 1990. The program funded research, not product development. Commercialization of the technology resulting from a project might overlap the research effort at a nascent level, but generally full translation of the technology into products and processes might take a number of additional years. ATP, through cost sharing with industry, invested in risky technologies that had the potential for spillover benefits to the economy.[2]

Appropriations to ATP increased from $10 million in 1990 to a peak of $341 million in 1995. Funding decreased in 1996 to $221 million, and it averaged about $200 million per year until 2004 when it fell to just under $150 million. The program was gradually phased out by 2007.

Through its intramural research program, ATP funded R&D projects performed internally by the NIST scientists in the Measurement and Standards Laboratories (MSLs) for development of infrastructure technology in support of ATP's mission. Under the statute governing ATP, up to 10% of ATP's budget could be allocated for intramural research within NIST. Beginning in 1997, ATP required that these intramural projects emphasize basic research, relate to groups of ATP extramural projects, and focus on measurement and standards that would facilitate the deployment and diffusion of ATP-funded technologies developed through ATP's extramural awards.

This chapter presents the findings from a non-economics-model-based evaluation of the outputs of ATP's intramural funding during the period from 1992 to 2000.

During the fiscal years 1992 through 2000, nearly $99 million was allocated toward intramural projects in the MSLs (see Fig. 6.1). Funding ranged from $4.7 million in fiscal year 1992 to almost $14 million in fiscal year 2000. Only in 1992, 1993, and 2000 did the allocation for intramural funding come close to the allowable 10% allocation. The allocation for the remaining years ranged from a low of 3% in 1995 to 7% in 1998 (see Table 6.1).

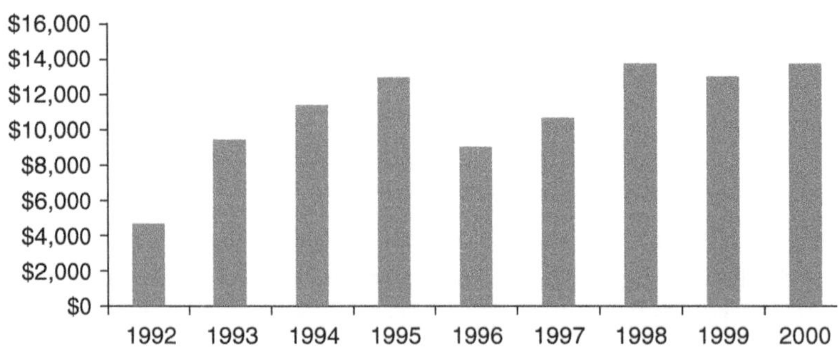

Figure 6.1 Allocation of intramural research funds ($1000s), by fiscal year.

Table 6.1 Intramural Research Funding as Percentage of ATP Allocations

Fiscal Year	Percentage (%)*
FY 1992	9.5
FY 1993	9.9
FY 1994	5.7
FY 1995	3.0
FY 1996	4.1
FY 1997	4.8
FY 1998	7.2
FY 1999	6.4
FY 2000	9.7

Note: The FY 1993 percentage does not include carryover funding from FY 1992.
*Percentages are rounded.
ATP, Advanced Technology Program; FY, fiscal year.

6.2. Laboratory Structure at NIST

The MSLs at NIST provide technical leadership for vital components of the nation's technology infrastructure needed by U.S. industry to improve continually its products and services. There are seven (with laboratory [lab] numbers as shown for use in the empirical models that follow) research laboratories at NIST:[3]

- The *Electronics and Electrical Engineering Laboratory* (EEEL, lab 81) promotes U.S. economic growth by providing measurement capability of high impact focused primarily on the critical needs of the U.S. electronics and electrical industries and their customers and suppliers.
- The *Manufacturing Engineering Laboratory* (MEL, lab 82) performs research and development of measurements, standards, and infrastructure technology related to manufacturing.
- The *Chemical Science and Technology Laboratory* (CSTL, lab 83) provides chemical measurement infrastructure to enhance U.S. industry's productivity and competitiveness, assure equity in trade, and improve public health, safety, and environmental quality.
- The *Physics Laboratory* (PL, lab 84) supports U.S. industry by providing measurement services and research for electronic, optical, and radiation technologies.
- The *Materials Science and Engineering Laboratory* (MSEL, lab 85) stimulates the more effective production and use of materials by

working with materials suppliers and users to assure the development and implementation of the measurement and standards infrastructure for materials.

- The *Building and Fire Research Laboratory* (BFRL, lab 86) enhances public safety and the competitiveness of U.S. industry by developing performance prediction methods, measurement technologies, and technical advances needed to assure the life-cycle quality and economy of constructed facilities.
- The *Information Technology Laboratory* (ITL, lab 89, including *Computer Systems*, lab 87) works with industry, research, and government organizations to develop and demonstrate tests, test methods, reference data, proof of concept implementations, and other infrastructural technologies.[4]

6.3. Allocation of Intramural Research Funds by Laboratory

Figure 6.2 shows the interlaboratory distribution of the ATP intramural research funds from 1992 through 2000. The MSEL received the largest share (almost one-fourth of the funding) during that time period. Three other laboratories, EEEL, MEL, and CSTL, each received about one-sixth

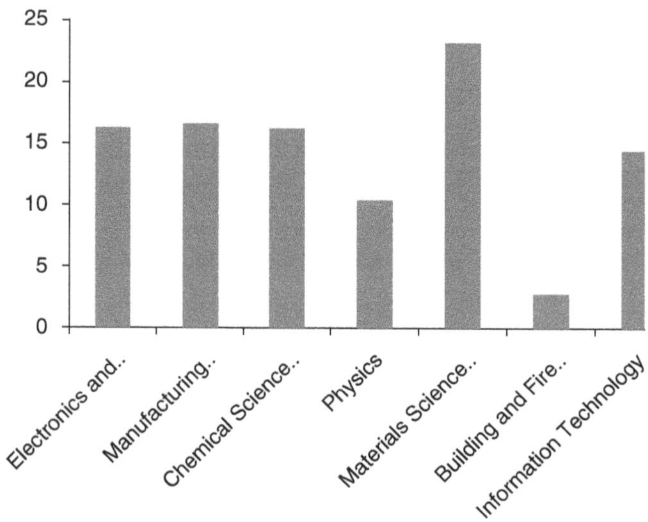

Figure 6.2 Percentage allocation of intramural research funds, by laboratory.

of the funding over the 9 years, and ITL received almost the same amount. The PL and the BFRL received the remaining funds.

6.4. Research Methodology and Data Collection Strategy

In an ideal experiment, one would compare the outputs and outcomes associated with research within NIST laboratories *with* ATP intramural funding to the outputs and outcomes associated with comparable NIST laboratories' research *without* ATP intramural funding. Another option would measure the outputs and outcomes associated with research in NIST laboratories over time and compare them between periods when ATP intramural funding was available and when it was not, all other things remaining constant. Neither approach was possible because comparable research projects could not be identified and because the relative newness of the intramural program (at the time of this study) made the time frame for a time series study too short to accommodate research projects starting and being completed.

Thus, as an alternative, NIST scientists who received ATP funding were asked to complete a survey that asked them to compare the outputs and outcomes from their research with hypothetically similar projects and broad NIST laboratory benchmarks. In addition, the outputs and outcomes from the ATP-funded projects were compared with the same type of outputs and outcomes for all of the NIST laboratories' projects.

In the fall of 2001, with input from the ATP, a survey instrument was designed for all ATP intramural project principal investigators (PIs). It was pretested and sent electronically to the 278 project PIs who were at NIST at that time.[5] During fiscal year 1992 through fiscal year 2000, ATP funded 1,052 intramural projects. The 278 surveyed PIs were associated with 510 of these projects.

An introductory letter from ATP's deputy director informed operating unit directors that the results of the survey were intended to help ATP know how to collaborate better with the laboratories using intramural funding. In addition, ATP expected statistics from the survey to be useful in summarizing some of the laboratories' achievements as a result of the funding.

In early December 2001, each PI who did not respond to the first survey was sent a reminder e-mail. No surveys were received after the end of 2001. A total of 209 surveys were returned with at least some

Table 6.2 PI Response Rates to the Survey

Laboratory	No. of PIs	Response Rate (%)*
Electronics and electrical engineering (81)	40	38
Manufacturing engineering (82)	38	58
Chemical science and technology (83)	54	32
Physics (84)	36	53
Material science and engineering (85)	57	40
Building and fire research (86)	13	38
Computer systems (87)	11	45
Information technology (89)	29	48
All	278	43

*Percentages are rounded.
PI, principal investigator.

information for one or more projects. This represents a response by 43% of the 278 PIs (see Table 6.2). The surveys included at least some information for 41% of the 510 projects;[6] 30 reported that no information was available on the project because of its age (see Table 6.3). Table 6.4 shows the project response rates by year.

Response rates differed for each question. Generally, the greater the number of intramural research awards that the PI received and the greater the age of the project, the less likely the PI was to respond to a question for a particular project. In contrast, the larger the budget of the project, the more likely the response.

These three response-determining factors are intuitive. The more projects for which a PI was asked information, the greater the total time

Table 6.3 Project Response Rates to the Survey

Laboratory	No. of Projects	Response Rate (%)*
Electronics and electrical engineering (81)	89	43
Manufacturing engineering (82)	72	54
Chemical science and technology (83)	85	25
Physics (84)	68	49
Material science and engineering (85)	115	43
Building and fire research (86)	18	28
Computer systems (87)	19	32
Information technology (89)	44	41
All	510	41

*Percentages are rounded.

Table 6.4 Response Rates by Fiscal Year of Project Origination

Fiscal Year	Response Rate*
FY 1992	16 of 41 projects = 39.0%
FY 1993	15 of 40 projects = 37.5%
FY 1994	22 of 52 projects = 42.3%
FY 1995	30 of 66 projects = 45.5%
FY 1996	28 of 73 projects = 38.4%
FY 1997	19 of 52 projects = 36.5%
FY 1998	32 of 75 projects = 42.7%
FY 1999	29 of 65 projects = 44.6%
FY 2000	18 of 46 projects = 39.1%
All	209 of 510 projects = 41%

*Percentages are rounded.
FY, fiscal year.

burden to the PI and the less likely he or she would respond. Similarly, greater time would be needed by the PI to retrieve requested information about an older project. PIs with larger budgets perhaps responded more often because the breadth and scope of the research was greater; hence, its importance to the PI's research agenda was greater. As explained subsequently, statistical adjustments were made for nonresponse bias.

6.5. Estimating the Probability of Survey Response

The response rates differed by question; therefore, a probability of response model was estimated for each individual question. The probability of response to a question is hypothesized to be a function of the difficulty of responding, the age of the project, the project's budget, and laboratory effects. The difficulty (*dif*) of responding for the PI is measured by the number of intramural research awards that the PI received.[7] The age (*age*) of the project is measured by the number of years since the project began. The total budget (*budget*) of the project is measured in constant 1996 dollars.[8] The laboratory (*lab81*,. . ., *lab89*) of the funded PI is also held constant in the model:

$$\text{Probability of response} = F\,(dif, age, budget, lab81, lab82, lab83, lab84, lab85, lab86, lab87, lab89) \tag{6.1}$$

When appropriate, the probit results from the probability of response model are provided throughout this chapter as each question is discussed. The response models show that not all explanatory variables were important for response to all of the questions. However, for those questions where the explanatory variables for response were important, and controlling for the other variables that affect response, a PI was less likely to respond to a question for a particular project the greater the difficulty of responding and the greater the age of the project. In contrast, other things being controlled, response was more likely the larger the project's budget. The laboratory effects on the response probability, held constant in each response model, are at times significant as well.

6.6. Project Outputs and Outcomes

Information was requested from each PI on broadly defined outputs and outcomes from each research project. These included the effect on the scope of laboratory research; counts of publications, citations, patents, presentations; and the leveraging effect on other funding sources. Each is discussed in order in the following sections.

6.6.1. Scope of Laboratory Research

Regarding the effect of intramural funding on the scope of laboratory research, each PI was asked about the relationship of the intramural project to ongoing research in the laboratory. As shown in Table 6.5, 56% ($n = 176$)[9] of the projects initiated a new research direction, that is, they expanded NIST competencies in new areas as opposed to complementing an ongoing research agenda.

Table 6.6 shows the joint estimation of the probability of responding to the scope question and the probability of answering "yes" to the question. Estimating jointly the probability of response and the probability that scope increased shows that the correlation of the disturbances for the two equations is positive and indeed almost equal to 1.0. When the error in the response model is large and response is more likely to occur, then the error in the probit model for scope is also large and scope is more likely to have increased. For that reason, after controlling for the response bias in the results, the percentages in Table 6.5 showing new research directions tend to be overestimates of the probability that the laboratories' intramural projects with the stated characteristics described

Table 6.5 Effect of ATP Intramural Funding on the Scope of Laboratory Research

Did the scope of this intramural research project complement an ongoing research agenda in your laboratory, or did it help to initiate a new research direction—that is, did it expand NIST's competencies in new areas? Yes = 1, No = 0

Laboratory	No. of Projects	Initiate New Research Direction(%)*
Electronics and electrical engineering (81)	25	76
Manufacturing engineering (82)	35	43
Chemical science and technology (83)	19	63
Physics (84)	29	55
Material science and engineering (85)	44	55
Building and fire research (86)	5	40
Computer systems (87)	5	60
Information technology (89)	14	57
All	176	56

*Percentages are rounded.
ATP, Advanced Technology Program; NIST, National Institute of Standards and Technology.

by the variables in the model increased the scope of the laboratory research.

Table 6.7 shows the response-adjusted probabilities as predicted by the maximum likelihood estimates of the probit model for scope and the probit model from response.[10] The model's estimates are shown in Table 6.6, and from that model the probability of a "yes" answer to the scope question is predicted for each observation in the complete 510 project sample.[11] The predictions are averaged for each laboratory to get the results shown in Table 6.7. On average, 20% of the projects are perceived to have an increase in scope. Perhaps reflecting the 1997 ATP policy shift of intramural funding emphasis to generic projects that could cut across a group of ATP projects, older projects are less likely to be perceived as increasing the scope of the laboratory's research.

6.6.2. Publications

Information was requested on the number of publications, either in print or accepted for publication and forthcoming that resulted directly from

Table 6.6 Probit Model of Scope with Sample Selection

Model	Variable	Coefficient	Standard Error
Scope	*age*	–0.06360**	0.02990
	budget	1.44×10^{-6} ***	3.93×10^{-7}
	lab82	–0.1503	0.2346
	lab83	–0.3794	0.2362
	lab84	0.01499	0.2291
	lab85	–0.1499	0.2054
	lab86	–0.4835	0.4270
	lab87	–0.07728	0.3868
	lab89	–0.2905	0.2808
	constant	–0.5794**	0.2367
Selection	*dif*	–0.07303**	0.03643
	age	–0.06444**	0.02646
	budget	1.65×10^{-6} ***	3.99×10^{-7}
	lab82	0.4847**	0.2118
	lab83	–0.3150	0.2169
	lab84	0.3584*	0.2120
	lab85	0.1915	0.1894
	lab86	–0.1358	0.3525
	lab87	0.003369	0.3465
	lab89	–0.1509	0.2567
	constant	–0.2003	0.2319
	athrho	4.513	56.87
	rho	0.9998	0.02738

No. of observations = 510
Censored observations = 334
Uncensored observations = 176
Log likelihood = –419.99
Wald $Chi^2(9)$ = 20.40
Probability > Chi^2 = 0.0156
LR test of independent equations (*rho* = 0): $Chi^2(1)$ = 3.07
Probability > Chi^2 = 0.0798

Notes: The results are for the joint maximum likelihood estimation of the probit model for scope and the response (selection) model for the 176 respondents that provided answers to the scope question. Coefficients for the variables in these and all subsequent probit models are the variables' coefficients in the probit index function. The statistic *athrho* is inverse hyperbolic tangent of *rho,* the correlation between the disturbances in the probit model for scope and in the selection model for response to the scope question. Significance (two-tails): ***significant at 0.01 level, **significant at 0.05 level, *significant at 0.10 level.

Table 6.7 Predicted Effect of ATP Intramural Funding on the Scope of Laboratory Research

The probability that the intramural research project initiated a new research direction as predicted by the model reported in Table 6.6: average of the probabilities predicted for each project in the categories shown.

Laboratory	No. of Projects	Probability Initiate New Research Direction(%)*
Electronics and electrical engineering (81)	89	22
Manufacturing engineering (82)	72	21
Chemical science and technology (83)	85	14
Physics (84)	68	24
Material science and engineering (85)	115	20
Building and fire research (86)	18	12
Computer systems (87)	19	17
Information technology (89)	44	19
All	510	20

*Percentages are rounded.
ATP, Advanced Technology Program.

each project. Of the 179 projects, the mean number of publications per project was 5.01; the range was 0 to 62 as shown in Table 6.8.[12] Also shown in Table 6.8 are the mean and range by laboratory.

The probit model for response to the question about publications is in Table 6.9, and the results show that the probability of responding decreases with difficulty and with age, but increases with budget. The model shown in Table 6.9 is also exactly the probit model for response to the questions about citations and about patents, because the respondents to all three questions were the same. Citations and patents are discussed subsequently.

A complete negative binomial maximum likelihood regression model (or count model) for the number of publications reported for each project can be readily estimated. However, when estimated as a single equation, one does not know whether the effects of the explanatory variables on the estimated number of publications reflect true explanatory effects for the population of ATP projects, or instead reflect the association of the variables with the probability of response, which is then in turn associated with the number of publications. To address this problem,

Table 6.8 Publications per Project Resulting from ATP Intramural Projects

Please provide complete bibliographic citations for all technical papers that resulted from this project and were published in professional journals, conference proceedings, or elsewhere.

Laboratory	No. of Projects	Mean No. of Publications	Range of Publications
Electronics and electrical engineering (81)	25	4.0	0–11
Manufacturing engineering (82)	35	3.7	1–17
Chemical science and technology (83)	19	6.7	0–61
Physics (84)	28	4.3	0–18
Material science and engineering (85)	47	7.8	0–62
Building and fire research (86)	5	1.8	0–4
Computer systems (87)	5	1.6	0–5
Information technology (89)	15	2.3	0–7
All	179	5.0	0–62

ATP, Advanced Technology Program.

the negative binomial model for publications was estimated simultaneously with the probit model for selection into the sample (that is, the model for the probability of response to the question about publications).[13] For some models it is very difficult to get maximum likelihood estimates of the response model and the count model of publications to converge; indeed, convergence and hence the estimates are available only for a very parsimonious count model for publications. That parsimonious model, estimated simultaneously with the full response model, is the only model of publications that is of interest or importance here. That is because we have the descriptive statistics for the reported publications as shown in Table 6.8. A fully specified single-equation negative binomial count model provides a good fit to those data for the 179 projects that responded, but we do not care about that model because it confounds response effects with the effects of underlying explanatory variables. If we want to have the descriptive statistics for the 179 reporting projects, we have Table 6.8, whereas if we want to estimate what the publications were for an ATP intramural project in the population, we

Table 6.9 Probit Model for Response to the Publications Question

Variable	Coefficient	Standard Error	Mean of X
constant	–2.3555***	0.7365	
dif	–0.07223*	0.04397	2.5882
age	–0.04449*	0.02714	5.7510
lnbudget	0.1999***	0.05962	11.4497
lab82	0.5094**	0.2098	0.1412
lab83	–0.3268	0.2177	0.1667
lab84	0.3112	0.2119	0.1333
lab85	0.2427	0.1876	0.2255
lab86	–0.1646	0.3544	0.03529
lab87	–0.06721	0.3471	0.03725
lab89	–0.07373	0.2533	0.08627

No. of observations = 510

Log likelihood function = –310.94

Restricted log likelihood = –330.51

Chi^2 = 39.13

Degrees of freedom = 10

Probability [Chi^2 > value] = 0.000024

Measures of fit: Estrella = 0.07604; McFadden = 0.05920

Note: Significance (two-tails): ***significant at 0.01 level, **significant at 0.05 level, *significant at 0.10 level.

need to use the best model that is estimable simultaneously with the model for response.

With correction for selection into the sample, Table 6.10 shows the negative binomial count model for publications that could be estimated with the data available for the sample of ATP intramural projects. The estimated number of publications increases with the natural logarithm of the budget, and laboratory 82 is expected to have fewer publications holding constant the effect of the budget (the model actually predicts slightly higher publications per project for the 72 projects in laboratory 82 once the effect of budget size is controlled). Simultaneously estimated with the negative binomial count model, the individual coefficients in the model of response are no longer significant, but they are in themselves not of interest to us. The maximum likelihood correlation coefficient between the error in the probit model of response and

Table 6.10 Negative Binomial Regression Model for the Number of Publications from a Project with Correction for Sample Selection

Variable	Coefficient	Standard Error	Mean of X
constant	–10.5813***	1.4514	
lnbudget	0.8635***	0.1126	11.69
lab82	–0.1590	0.2735	0.1955
Overdispersion Parameter for Negative Binomial			
theta	0.2272*	0.1215	
Standard Deviation of Heterogeneity			
sigma	1.5611***	0.2014	
Correlation of Heterogeneity and Selection			
rho	1.000	0.6184	
No. of observations = 510			
Log likelihood function = –746.30			
Restricted log likelihood =–1210.22			
Chi² = 927.83 with 2 degrees of freedom			
Probability [Chi² > value] = 0.0000			
Mean of LHS variable = 5.006			
Restricted Log-likelihood is Poisson + probit (independent)			
Log likelihood for initial probit = –310.94			
Log likelihood for initial Poisson = –899.28			
Means for Poisson/negative binomial use selected data			

Notes: Significance (two-tails): ***significant at 0.01 level, **significant at 0.05 level, *significant at 0.10 level.

The customary overdispersion parameter *alpha* has been set equal to 1/*theta* to simplify the numerical computations in the maximum likelihood problem. The value reported for *theta* shows it is clearly less than 1; hence, *alpha* is clearly greater than 1, and there is overdispersion, appropriately treated with the negative binomial generalization of the Poisson. Alternatively, the negative binomial model can be interpreted as the result of introducing unobserved individual heterogeneity into the Poisson model. For technical description and details, see Greene (2002b, Chapter E20; 2003, pp. 740–745).

in the count model is almost 1.0. From the estimated coefficients in the negative binomial model as corrected for sample selection, we have the predicted number of publications for a project equal to *e*, the base of the natural logarithms, raised to the power (–10.58 + 0.8635 *lnbudget* – 0.1590 *lab82*).

Following this formulation, Table 6.11 uses the negative binomial model (eliminating the selection effect) to predict the number of publications for

Table 6.11 Predicted Publications per Project Resulting from ATP Intramural Projects

Laboratory	No. of Projects	Mean No. of Publications*	Range of Publications*
Manufacturing engineering (82)	72	0.7	0.05–4.9
All except manufacturing engineering (82)	438	0.7	0.3–6.7
All	510	0.7	0.05–6.7

*Numbers are rounded.
ATP, Advanced Technology Program.

each of the 510 projects in the complete sample. Table 6.11 shows the average predictions, which, because of the elimination of the selection bias, show fewer predicted publications as compared to the actual numbers of publications per project for those responding to the survey. The model's prediction is that on average for all of the 510 projects, there were about two publications for every three projects (0.69 publications per project), rather than five publications per project as suggested by the respondents to the survey. To provide an indication of the robustness of the prediction here, we also estimated the negative binomial model with control for sample selection by using a two-step method—estimating a nonselection hazard rate from the probit model of response and then including it as a variable in the negative binomial estimation. The two-step method does not face the difficulties of obtaining convergence that occur with the full information maximum likelihood estimation of the selection and count models simultaneously. As a result, all of the explanatory variables other than the identifying variable, *dif*, could be included in the count model. Instead of predicting 0.69 publications per project, the two-step model predicts 0.65 publications per project—essentially the same answer of about two publications for every three projects.[14]

6.6.3. Citations

The number of citations of each in-print publication was calculated on the basis of information in the Expanded Science Citation Index, as accessed through the Web of Science. Each publication that a PI reported on his or her returned survey was verified in the Expanded Science

Citation Index and the number of non–self-citations (including all co-authors) was counted. The mean (n = 133) number of non–self-citations per publication for those projects reporting publications was 1.66; the range was 0 to 33 (see Table 6.12).

The probit model for response to the citations question is identical to the model for response in Table 6.9, because the respondents to the two questions were the same. Table 6.13 shows the negative binomial model for the counts of citations per project with correction for sample selection. The procedure used is the same as described in the section about the count model of publications.

The model was used to predict the number of citations for each of the 510 projects, and taking that prediction and then dividing by the number of publications for each project, Table 6.14 shows the average predictions per publication for the typical project with the characteristics of the 133 projects reporting publications. The predicted number of citations per project is e raised to the power (–11.18 + 0.4695 *age* + 0.4560 *lnbudget* + 3.40 *lab83* + 3.96 *lab84* + 2.43 *lab85*).

For all projects having publications, the average number of citations is predicted to be about the same as the one and two-thirds citations per publication for the responding projects. Typical projects would have fewer publications and fewer citations, but the citations per publication would be about the same as for the reporting projects.

Table 6.12 Citations per In-Print Publication Resulting from ATP Intramural Projects

Laboratory	No. of Projects	Mean No. of Citations*	Range of Citations*
Electronics and electrical engineering (81)	18	0.2	1–1.3
Manufacturing engineering (82)	23	0.1	0–1.6
Chemical science and technology (83)	15	1.2	0–4.6
Physics (84)	21	5.2	0–33
Material science and engineering (85)	39	2.3	0–32.7
Building and fire research (86)	3	0	–
Computer systems (87)	3	0.2	0–0.5
Information technology (89)	11	0.1	0–0.5
All	133	1.7	0–33

*Numbers are rounded.
ATP, Advanced Technology Program.

Table 6.13 Negative Binomial Regression Model for the Number of Citations per Project with Correction for Sample Selection

Model	Variable	Coefficient	Standard Error	Mean of X
Negative binomial	constant	–11.1785***	1.5215	
	age	0.4695***	0.05978	5.4022
	lnbudget	0.4560***	0.1062	11.6859
	lab83	3.4042***	0.4749	0.1061
	lab84	3.9578***	0.4772	0.1564
	lab85	2.4298***	0.4204	0.2626
Selection	constant	–2.3697***	0.7311	
	dif	–0.07404	0.05028	2.5882
	age	–0.04296	0.02898	5.7510
	lnbudget	0.2001***	0.05943	11.4497
	lab82	0.5248**	0.2119	0.1412
	lab83	–0.3151	0.2317	0.1667
	lab84	0.3240	0.2125	0.1333
	lab85	0.2433	0.1885	0.2255
	lab86	–0.1455	0.3511	0.03529
	lab87	–0.05030	0.3488	0.03725
	lab89	–0.06848	0.2595	0.08627
Overdispersion Parameter for Negative Binomial				
	theta	0.08947**	0.03696	
Standard Deviation of Heterogeneity				
	sigma	2.1222***	0.1493	
Correlation of Heterogeneity and Selection				
	rho	0.1249	0.1456	

No. of observations = 510
Log likelihood function = –633.51
Restricted log likelihood = –2713.54
Chi^2 = 4160.06 with 2 degrees of freedom
Probability[Chi^2 > value] = 0.0000
Mean of LHS variable = 10.29
Restricted log-likelihood is Poisson + probit (independent)
Log likelihood for initial probit = –310.94
Log likelihood for initial Poisson = –2402.60
Means for Poisson/negative binomial use selected data

Notes: Significance (two-tails): ***significant at 0.01 level, **significant at 0.05 level, *significant at 0.10 level.
The customary overdispersion parameter *alpha* has been set equal to 1/*theta* to simplify the numerical computations in the maximum likelihood problem. The value reported for *theta* shows it is clearly less than one; hence, *alpha* is clearly greater than one and there is over-dispersion, appropriately treated with the negative binomial generalization of the Poisson. Alternatively, the negative binomial model can be interpreted as the result of introducing unobserved individual heterogeneity into the Poisson model. For technical description and details, see Greene (2002b, Chapter E20; 2003, pp. 740–745).

Table 6.14 Predicted Citations per Predicted Publication from ATP Intramural Projects, with Correction for Selection, for the 133 Projects Reporting at Least One Publication

Laboratory	No. of Projects	Mean No. of Citations per Publication*	Range of Citations per Publication*
All	133	1.8	0.01–22.9

Note: The count model for publications with correction for sample selection predicts an average of 1.00 publications for the 133 projects for which actual publications were 5.01 per project (Table 6.8). The error in the selection model was highly correlated with the error in the publications count model, so the 133 reporting projects had more publications than expected for the typical projects with their characteristics. The count model for citations with correction for sample selection predicts an average of 1.48 citations for the 133 projects. The ratio of the predicted citations to the predicted publications averaged across the 133 projects is 1.76. The reporting projects had more publications than expected for the typical project and somewhat more citations; the ratio, or citations per publication, for the typical project with publications is essentially the same as for the 133 reporting projects.
*Numbers are rounded.
ATP, Advanced Technology Program.

6.6.4. Patents

Patenting (at the time of this study) was not an integral part of the research mission of NIST scientists. However, a few PIs for selected projects did report that a patent had either been filed or had been received. As shown in Table 6.15, on average (n = 179), the number of patents per project was 0.05 with a range of 0 to 3. Disaggregation of the survey data by laboratory is not meaningful for this output measure because so few patents had been filed or received; for the same reason, estimating a count model for patents is not sensible.

6.6.5. Presentations

Presentations are the primary mode for quickly and effectively disseminating information from NIST research. On average (n = 174), the number of presentations per project was 9.55 with a range of 0 to 150 (see Table 6.16).

The probit model for response to the presentations question is shown in Table 6.17. Table 6.18 shows the negative binomial model estimated simultaneously with the model for response following the procedure introduced to model publications. The model was used to predict the

Table 6.15 Patents per Project from the ATP Intramural Projects

Laboratory	No. of Projects Reporting	Mean No. of Patents*	Range of Patents
All	179	0.05	0–3

*Number is rounded.
ATP, Advanced Technology Program.

number of presentations for each of the 510 projects as *e* raised to the power (–11.32 + 0.9177 *lnbudget* + 0.5885 *lab82* + 0.5774 *lab83* + 1.007 *lab84* + 0.1414 *lab85* + 0.1303 *lab86* – 1.013 *lab87* + 0.3731 *lab89*).

Table 6.19 shows the average prediction per project for each laboratory after correcting for the response effect. About 1 presentation per project is expected on average for all 510 projects, rather than almost 10 presentations per project for those responding to the survey.[15]

Table 6.16 Presentations per Project Resulting from ATP Intramural Projects

Approximately, how many professional presentations (outside of NIST) related to this project have been given?

Laboratory	No. of Projects	Mean No. of Presentations*	Range of Presentations
Electronics and electrical engineering (81)	25	4.2	0–15
Manufacturing engineering (82)	34	7.9	0–60
Chemical science and technology (83)	19	17.3	0–150
Physics (84)	28	7.6	0–40
Material science and engineering (85)	43	15.4	0–150
Building and fire research (86)	5	3.4	0–10
Computer systems (87)	5	2.6	0–10
Information technology (89)	15	3.4	0–9
All	174	9.6	0–150

*Numbers are rounded.
ATP, Advanced Technology Program; NIST, National Institute of Standards and Technology.

Table 6.17 Probit Model of Response to Presentations Question

Variable	Coefficient	Standard Error	Mean of X
constant	–2.281***	0.7388	
dif	–0.09528**	0.04459	2.588
age	–0.04515*	0.02721	5.751
lnbudget	0.2000***	0.05993	11.45
lab82	0.4639**	0.2099	0.1412
lab83	–0.3531	0.2180	0.1667
lab84	0.3026	0.2120	0.1333
lab85	0.1392	0.1889	0.2255
lab86	–0.1956	0.3550	0.03529
lab87	–0.08796	0.3479	0.03725
lab89	–0.1013	0.2539	0.08627

No. of observations = 510
Log likelihood function = –307.91
Restricted log likelihood = –327.32
Chi^2 = 38.84 with 10 degrees of freedom
Probability [Chi^2 > value] = 0.2710×10^{-4}
Fit Measures: Estrella = 0.07550; McFadden = 0.05933

Note: Significance (two-tails): ***significant at 0.01 level, **significant at 0.05 level, *significant at 0.10 level.

6.6.7. Leveraging Additional Funding

In addition to the knowledge spillovers associated with publications, citations, and presentations, the ATP intramural projects also leveraged other sources of funding such as NIST competency awards or CRADA activity.[16] For the responding projects (n = 176), 31% of the projects leveraged additional funding as shown in Table 6.20.

Table 6.21 presents the maximum likelihood estimation of the probit model with sample selection.[17] Although the model of selection (response) is quite significant and tells the same story we have already seen about response, there are not significant differences across the labs in the probability of leveraging subsequent competency awards. The differences not being significant, we report for the 510 projects in the complete sample simply the average prediction of that probability for a project. That average probability is 0.14, and the range across the 510 projects for the predicted probability of leveraging is from 0.02 to 0.37. However, we cannot reject the possibility that the selection and leveraging models are

Table 6.18 Negative Binomial Regression Model for the Number of Presentations per Project with Correction for Sample Selection

Variable	Coefficient	Standard Error	Mean of X
constant	−11.32***	0.8405	
lnbudget	0.9177***	0.06440	11.69
lab82	0.5885***	0.2192	0.1954
lab83	0.5774***	0.2163	0.1092
lab84	1.007***	0.2592	0.1609
lab85	0.1414	0.2058	0.2471
lab86	0.1303	0.4035	0.02874
lab87	−1.013	0.9033	0.02874
lab89	0.3731	0.3363	0.08621
Overdispersion Parameter for Negative Binomial			
theta	0.05007***	0.01794	
Standard Deviation of Heterogeneity			
sigma	1.619***	0.07094	
Correlation of Heterogeneity and Selection			
rho	1.000***	0.1166	

No. of observations = 510
Log likelihood function = −795.80
Restricted log likelihood = −2030.52
Chi^2 = 2469.43 with 2 degrees of freedom
Probability [Chi^2 > value] = 0.0000
Mean of LHS Variable = 9.534
Restricted Log-likelihood is Poisson + probit (independent)
Log likelihood for initial probit = −307.91
Log likelihood for initial Poisson = −1722.62
Means for Poisson/negative binomial use selected data

Notes: Significance (two-tails): ***significant at 0.01 level, **significant at 0.05 level, *significant at 0.10 level.
The customary overdispersion parameter *alpha* has been set equal to 1/*theta* to simplify the numerical computations in the maximum likelihood problem. The value reported for *theta* shows it is clearly less than 1; hence, *alpha* is clearly greater than 1, and there is overdispersion, appropriately treated with the negative binomial generalization of the Poisson. Alternatively, the negative binomial model can be interpreted as the result of introducing unobserved individual heterogeneity into the Poisson model. For technical description and details, see Greene (2002b, Chapter E20; 2003, pp. 740–745).

Table 6.19 Predicted Presentations per Project Resulting from ATP Intramural Projects

Laboratory	No. of Projects	Mean No. of Presentations*	Range of Presentations*
All	510	1	0.02–10.6

*Numbers are rounded.
ATP, Advanced Technology Program.

independent, and so perhaps the estimates from our respondents in Table 6.20 are the best ones.

6.7. NIST Benchmarks

To place these outputs and impacts in perspective, ATP provided selected NIST-wide benchmark data. As shown in Table 6.22, the performance of the ATP-funded projects that reported their publications in response to the survey had a much higher number of publications per million dollars of budget (in constant 1996 dollars) than the average project in the same NIST laboratory. The estimated models of response and of publications

Table 6.20 Impact of ATP Intramural Project on Leveraging Other Sources of Funding

Did this project contribute to the basis for subsequent NIST competency awards, other sources of federal funding to the laboratory, or CRADA activity? Yes = 1, No = 0

Laboratory	No. of Projects	Leveraging Rate(%)*
Electronics and electrical engineering (81)	25	24
Manufacturing engineering (82)	35	20
Chemical science and technology (83)	18	39
Physics (84)	27	44
Material science and engineering (85)	47	34
Building and fire research (86)	5	20
Computer systems (87)	5	20
Information technology (89)	14	29
All	176	31

*Numbers are rounded.
ATP, Advanced Technology Program; NIST, National Institute of Standards and Technology.

Table 6.21 Probit Model of Leveraging Competency Awards or Similar Funded Awards with Sample Selection

Model	Variable	Coefficient	Standard Error
Competency	*age*	0.03821	0.06000
	lnbudget	0.2448**	0.1021
	lab82	–0.001761	0.3903
	lab83	0.1585	0.4583
	lab84	0.4983	0.3429
	lab85	0.3291	0.3013
	lab86	–0.1419	0.6127
	lab87	–0.3292	0.6559
	lab89	0.1645	0.4259
	constant	–4.3372***	1.3162
Selection	*dif*	–0.06362	0.04311
	age	–0.06627**	0.02703
	budget	1.60×10^{-6}***	4.01×10^{-7}
	lab82	0.4850**	0.2129
	lab83	–0.3567	0.2214
	lab84	0.2831	0.2150
	lab85	0.2592	0.1892
	lab86	–0.1488	0.3539
	lab87	0.01510	0.3485
	lab89	–0.1433	0.2568
	constant	–0.2059	0.2388
	athrho	0.7822	1.099
	rho	0.6540	0.6292

No. of observations = 510
Censored observations = 334
Uncensored observations = 176
Log likelihood = –407.64
Wald $Chi^2(9)$= 12.66
Probability > Chi^2 = 0.1788
LR test of independent equations (*rho* = 0): $Chi^2(1)$ = 0.68 Probability > Chi^2 = 0.4107

Notes: Significance (two-tails): ***significant at 0.01 level, **significant at 0.05 level, *significant at 0.10 level.
The statistic *athrho* is inverse hyperbolic tangent of *rho* the correlation between the disturbances in the probit model for scope and in the selection model for response to the scope question; it is used to derive the estimate of *rho*.

Table 6.22 Comparison of Publications for ATP Intramural and NIST Projects

ATP Intramural Projects with Starting Dates from FY94 through FY00 and Respondents with Publications and Budget Data

Laboratory	*n*	Publications	Budget (1996 $K*)	Publications per $M of Budget*
Electronic and electrical engineering (81)	24	**93**	2,983	31.2
Manufacturing engineering (82)	29	**100**	7,015	14.3
Chemical science and technology (83)	17	**67**	4,223	15.9
Physics (84)	20	**72**	3,623	19.9
Material science and engineering (85)	44	**352**	6,613	53.2
Building and fire research (86)	4	**9**	644	14.0
Information technology (89)	17	**38**	2,771	14.0

All NIST Laboratories from FY94 through FY00

Laboratory	Average Annual Publications	Average Annual Budget (1996 $K*)	Publications per $M of Budget*
Electronic and electrical engineering (81)	232	36,884	6.3
Manufacturing engineering (82)	163	28,779	5.7
Chemical science and technology (83)	350	41,127	8.5
Physics (84)	276	29,920	9.2
Material science and engineering (85)	489	41,792	11.7
Building and fire research (86)	241	20,856	11.6
Information technology (89)	197	40,703	4.8

Note: The 510 observation sample of ATP intramural projects includes projects as old as those beginning in FY 1992 (with age 2002–1992 = 10) and as young as those beginning in FY00 (with age 2002–2000 = 2). NIST provided budget data for each lab for FY94 through FY01, as well as publications data by lab for those years and the years before and FY02. To have comparable data, in the table we used *(1)* ATP projects between 2 and 8 years old (the youngest starting in FY00 and the oldest starting in FY94), and *(2)* the NIST lab data for FY 1994 though FY 2000.
As the estimated models (both the approach as described in endnote 13 and the ad hoc two-step approach) show, age is not, in the sample, a factor explaining publications where the youngest projects are 2 years old. For applied science and technology, the papers come fairly quickly. The publications and budgets for the ATP intramural projects are observed at the project level, and the matching is therefore exact. For the NIST lab performance, the publications and budget data are at the lab level. The appropriate lag between project funding and publications is not known; furthermore, the projects extend over multiple years with publications coming throughout the projects' lives. Examination of the annual publications and budget (in constant 1996 dollars) data suggest a stable, annual relation between publications and budget, so for the NIST performance results the average annual number of publications per million dollars of budget was used.
*Numbers are rounded.
K denotes thousands; M denotes millions.
ATP, Advanced Technology Program; FY, fiscal year; NIST, National Institute of Standards and Technology.

do suggest that the responding ATP-funded projects performed exceptionally well relative to all ATP intramural projects. If the predictions of the model of publications with control for selection effects were used instead of the actual performance of the ATP-funded projects that responded to the survey, the predicted publications per million dollars of budget for the ATP intramural projects would not outperform the typical NIST projects. The predictions, however, are not the actual performance observed, but rather would be the best guess about the performance of an analogous, hypothetical project apart from random error. On average, the actual performance regarding publications for the reporting ATP intramural projects far exceeded the performance of the typical NIST laboratory projects.

6.8. Hypothetical Effects of Intramural Funding

In addition to documenting outputs—publications, citations, patents, presentations—in the non-economics-based evaluation models, the PIs were asked hypothetical questions to compare the nature of the intramural funded project to research activities that would have occurred in the absence of such funding. Detailed statistics are presented in Link and Scott (2004a, 2005a), and an overview of the findings is reported there.

Overall ($n = 151$), PIs reported that 46% of the projects would have been undertaken as a counterfactual benchmark project with similar goals and milestones had ATP funding not been received. However, these hypothetical projects, although with similar goals and milestones, would have taken longer to complete, would have been less technically challenging, and would have generated fewer technical outputs, as will be discussed later.

As a result of ATP funding, 84% of the research that was conducted was ahead by an average of 12 months ($n = 69$) as compared to where the research would have been had a project with similar goals and milestones been undertaken absent ATP intramural funding. With control for response bias, essentially all the labs show for their ATP-funded projects a probability of being ahead that is very close to 1.0.

Also, we examined several additional performance measures. On average ($n = 69$), 49% of the ATP intramural projects reported that they were broader in scope than the hypothetical projects. On average ($n = 66$), 58% of the ATP intramural projects were more technically challenging than the hypothetical projects. On average (n = 67), the expected durations of 55% of the ATP intramural projects are shorter than for the hypothetical projects. Furthermore, ATP intramural projects, in comparison to the hypothetical projects with similar goals and milestones that would have been undertaken without ATP funding, are expected to lead to more technical papers, more new measurement technology, more or at least the same number of new standards, and the same number of new databases.

Of the responding ATP intramural projects, 71% are expected to lead to more technical papers than what would be expected from the similar hypothetical projects without ATP support, and 64% are expected to lead to more new measurement technology. Only 3% of the ATP intramural projects are expected to lead to fewer new standards, meaning that 97% are expected to lead to more or at least the same number of new standards. Only 4% of the responding ATP intramural projects expected fewer new databases in comparison to the hypothetical projects, meaning that 96% are expected to lead to more or at least the same number of new databases. PIs were similarly asked about patents; the responding projects show that their numbers of patents would be about the same, which is close to zero as noted previously.

PIs responded that for 70 (46%) of 151 projects they would have undertaken a project with similar goals and milestones in the absence of funding for their ATP intramural project. For those 81 projects that would not have done so, PIs were asked if they would have alternatively undertaken a project in the same broadly defined research area. They report that, absent ATP funding, for 40% of the projects, alternative projects would have been undertaken in the same broadly defined research area. The total number of projects that fall within this category is small, but the responses report the following relative performance for those projects that would not have been undertaken with similar goals and milestones but would alternatively have been undertaken in the same broadly defined research area. As compared to the counterfactual project in the same broad area but without ATP support, the ATP-supported projects report that 52% are expected to be broader in scope; 48% are expected to be more technically challenging; 39% are expected to be of shorter duration; 44% are expected to generate more technical papers; 57% are expected to generate more new measurement technology; 53% are expected to generate more standards; and 69% are expected to generate more or at least the same number of new databases.

6.9. Conclusions

Less than one-half of the intramurally funded projects would have been undertaken in the absence of ATP funding. That finding alone supports, from an evaluation perspective, the leveraging impact of ATP's intramural research initiative. However, those that would have been undertaken, with similar goals and milestones, would have taken longer to complete, would have been less technically challenging, and would have generated fewer technical outputs. Although such a finding does not itself (that is, without developing the associated effects on value and cost) translate into traditional evaluation metrics, it is nonetheless valuable descriptive information for characterizing the impact of intramural funding.

The non-economics-based methodology presented in this chapter can be generalized for use in evaluations of programs and projects that differ from the publicly funded, publicly performed intramural research programs. Clearly there is a direct application of our ATP intramural analysis to other publicly funded, publicly performed programs that are also intramural. Table 6.23 identifies other prominent examples of such intramural programs that could be evaluated using our methodology. Table 6.24 illustrates the analogy of other intramural evaluations to the one that we have carried out for NIST and ATP. To NIST's in-house research that was supported intramurally by ATP (the prominent NIST extramural-research program during the period we have studied in this chapter), we have compared counterfactual, similar, in-house NIST

Table 6.23 Federal R&D Obligations to Intramural Activity in Selected Agencies, 2003 ($millions)

Agency	Total R&D Obligations	Intramural
All federal government	$98,608	$24,558
Department of Defense	$45,012	$12,408
Department of Health and Human Services	$27,551	$5,162
National Aeronautics and Space Administration	$8,598	$2,150
Department of Energy	$7,541	$764
Department of Commerce	$1,065	$838
Environmental Protection Agency	$627	$284

Source: National Science Board (2004).

Table 6.24 Application of the Methodology to Publicly Funded, Publicly Performed Intramural Research Programs

Program to Be Evaluated	The Counterfactual
NIST research with ATP support	NIST research without ATP support
Other intramural research programs	In-house research programs without intramural support

ATP, Advanced Technology Program; NIST, *National Institute of Standards and Technology.*

research without intramural support. The in-house research of other government agencies shown in Table 6.23 that is supported by the extramural programs of those agencies could also be compared with counterfactual, similar, in-house research without the intramural support.

Of course, the comparison of nonmarket performance measures for any program or project that combines outside resources with an organization's own resources to a counterfactual case without outside resources can be a useful evaluation method even beyond the publicly funded, publicly performed intramural projects. We expect that the approach would be especially useful for evaluating the contribution of outside assistance for within-country economic development projects in developing areas of a country or in developing nations where markets—and hence market values—may not be well established. For example, if a rural area of a developed country invited a national development program to work with local authorities to plan and implement regional development, our method could be used to evaluate the value added of the outside assistance. Likewise, if a developing nation invited the World Bank, for example, to send a team of experts to work with the nation's government to plan and implement a development program, our method provides a systematic way to evaluate the benefits gained from having the assistance of the World Bank's development team, as contrasted with working solely with the nation's internal resources (assuming that the program would have been undertaken).

Whether the programs and projects are public or private, or a mixture of the two, evaluation of the benefits of partnering could use our approach. The approach may be the primary one when information about market benefits would not typically be available. But the approach would be a useful addition to the collection of evaluation tools even when market data or estimated data are available.

The nonmarket performance indicators are of interest in themselves. Thus, the method proffered in this chapter would be useful for any investment programs that bolster the effort of one organization—analogous to NIST laboratories in our analysis—with the help of another organization—analogous to ATP in our analysis—and for which we want to augment conventional evaluation methods with the counterfactual comparison for nonmarket performance indicators for more complete evaluation. ATP's connections with industry and the monitoring and guidance in the context of the connection to industry provide valuable input to the work in the NIST labs, and as a result the performance of the projects is better as compared to the counterfactual case without the ATP support. By analogy, the outside partner in any program or project brings the diversity of another perspective and other types of resources into the work and improves the performance of the primary partner. Table 6.25 illustrates the program being evaluated and the counterfactual case.

In the most general sense, there is no need to restrict the analogy to other publicly financed, publicly performed programs and projects. The analogy and hence our nonmarket counterfactual method could apply to privately financed and performed projects and to mixtures of private and public finance and performance, too. The nonmarket counterfactual evaluation allows assessment of the impact of outside partners on program performance when market information is unavailable, and it provides useful additional information for assessment even when market valuations are available.

Finally, although the non-economics-based outputs and outcomes presented herein do not directly translate into traditional evaluation metrics—internal rates of return, benefit–cost ratios, and net present value estimates—our alternative measures nevertheless provide systematic information collected through a counterfactual experiment about the outputs and outcomes of the program in a way that embraces the spirit of public accountability.

Table 6.25 Application of the Methodology to Research Programs in General

Program to Be Evaluated	The Counterfactual
Primary partner/outside partner	Primary partner's project without outside partner's support

Notes

1. This chapter draws from Link and Scott (2004a; 2005a; 2009). Additional details are provided in those documents.

2. ATP provided incentives to firms to undertake research that would not otherwise have been pursued—like project A in Figure 2.1 in Chapter 2.

3. In addition to these research laboratories, Technology Services provides a variety of products and services to U.S. industry such as standard reference materials, standard reference data, and weights and measures. There are also two research centers at NIST, The Center for Nanosacale Science and Technology and the NIST Center for Neuron Research.

4. The Computer Systems Laboratory (CSL) and the Computing and Applied Mathematics Laboratory (CAML) were combined on February 16, 1997 to form the Information Technology Laboratory. For purposes of interlaboratory comparisons, pre-1997 information on these two MSLs is reported under the Information Technology Laboratory.

5. The survey instrument is provided in Link and Scott (2004a). The PI for an intramural project was identified to be that individual listed on the ATP intramural project budget allocation records at the time the award was made. If a project was awarded to more than one researcher, it was assumed that the first named researcher was the PI.

6. The survey had several open-ended questions, but very few PIs responded to them and thus no discussion of those sparse responses is presented herein.

7. The number of awards is also the number of survey instruments that the PI was asked to complete, and that number is expected to measure the difficulty that the PI faced in responding to the survey.

8. The budget figures were converted to constant 1996 dollars using Council of Economic Advisers (2002), Table B7, "Chain-type price indexes for gross domestic product, 1959-2001." The index numbers, 1996 = 100, for gross domestic product were used.

9. Not all questions were answered on all returned surveys.

10. The probabilities are predicted from the maximum likelihood probit model with selection—"heckprob" as described in StataCorp (2001) Vol. 2, pp. 31–39.

11. The predictions are made using the "pmargin" option for predictions (StataCorp 2001, Vol. 7, p. 31).

12. A priori, one might expect that pre-1997 projects' publication rates would differ from the publication rates for projects in subsequent years. The focus of the intramural awards program changed in 1997 from supporting specific external projects to supporting more broad-based projects. While the 179 reporting projects averaged 5.01 publications each with standard deviation 9.2 and range from 0 to 62, the 82 pre-1997 projects averaged 6.1 publications each with standard deviation 11.6 and a range from 0 to 62, and the 97 reporting projects from 1997 onward averaged 4.1 publications each with standard deviation 6.4 and a range from 0 to 46. However, in the model of publications with control for other variables and for sample selection, the publication rates for the older projects and the more recent ones are essentially the same.

13. The version of LIMDEP that we used included the program to estimate models for count data with a correction for sample selection; see Greene (2002a), "What's New in Version 8.0? WN2.7. Models for Count Data," *Reference*

Guide, p. WN-9: "The Poisson and negative binomial models can be fit with a correction for sample selection. . . . Estimation of the selection model is done by full information maximum likelihood. . . ." We use the full information maximum likelihood model for the negative binomial models with selection that are estimated in this chapter.

14. Our alternative approach is less orthodox than the formal approach using Greene's (2002a; 2002b; 2003) full information maximum likelihood approach described in the preceding note, and we present only the results of the formal approach here.

15. Again, it is worth noting that the result is the same result we obtained initially using a two-step estimator controlling for the nonselection hazard rate. That model predicted 0.90 presentations per project as compared to the prediction here of 1.01—essentially one presentation per project.

16. A Cooperative Research and Development Agreement (CRADA) is a written agreement between federal and nonfederal parties to cooperate on a research project.

17. See StataCorp. (2001), Vol. 2, pp. 31–39.

7

Wavelength References for Optical Fiber Communications

7.1. Background Information

The Optoelectronics Division of the Electronics and Electrical Engineering Laboratory (EEEL) at the National Institute of Standards and Technology (NIST) began research on optical communications in the mid-1970s and expanded its research program substantially in the late 1980s.[1] The Optical Fiber and Components Group of the Division began research on standard reference materials (SRMs) in 1991. The group's first SRM became available in 1993 with SRM 2520, an optical fiber diameter standard. Since then the group has produced a number of optoelectronic standards. SRM 2517 was issued in 1997; it was intended for use in calibrating the wavelength scale of wavelength-measuring equipment in the spectral region from 1,510 nm to 1,541 nm.

In 1998, Dr. Sarah Gilbert in the Optical Fiber and Components Group began a 2-year ATP-sponsored intramural project to develop a more accurate version of SRM 2517. She received $145,000 over 2 years—$70,000 in fiscal year 1998 and $75,000 in fiscal year 1999.

The project produced the new SRM for calibration of wavelengths in the spectral region from 1,510 nm to 1,540 nm. The references in the 1,500 nm region are important to support wavelength division multiplexed (WDM) optical fiber communications systems. In a WDM system, many channels, each associated with a different wavelength, of communications information are sent down the same fiber. Thus, wavelength division multiplexing in effect increases the bandwidth of the communications system because any given spectral region will support more channels through which communications information can be sent. A WDM system requires stable wavelengths throughout the components of the system, and equipment must be calibrated to measure those wavelengths. The wavelength references provided by NIST are needed to calibrate the instruments—such as optical spectrum analyzers, tunable lasers, and wavelength meters—that are used to characterize the components of WDM optical fiber communications systems. The wavelength references are also used to monitor the wavelengths of the channels while the system is in use; if one channel's wavelength were to shift, crosstalk could occur between it and a neighboring channel, thus disrupting the accurate flow of communications information through the channels of the system.

The output of Dr. Gilbert's Advanced Technology Program (ATP)-funded NIST research with William Swann was SRM 2517a, High Resolution Wavelength Calibration Reference for 1,510 nm to 1,540 nm Acetylene ($^{12}C_2H_2$). NIST's description of the new SRM provides an exact description of the artifact—an absorption cell filled with acetylene gas that produces characteristic absorption lines in the readouts resulting when lasers project light of various wavelengths through the gas-filled cell. The absorption lines observed can then be used to identify the wavelengths for the laser emitting device being calibrated.[2] NIST's description of the artifact is as follows:

> Standard Reference Material 2517a is intended for wavelength calibration in the spectral region from 1510 nm to 1540 nm. It is a single-mode optical-fiber-coupled absorption cell containing acetylene ($^{12}C_2H_2$) gas at a pressure of 6.7 kPa (50 Torr). The absorption path length is 5 cm and the absorption lines are about 7 pm wide. The cell is packaged in a small instrument box (approximately 24 cm long x 12.5 cm wide x 9 cm high) with two FC/PC fiber connectors for the input and output of a user-supplied light source. Acetylene has more than 50 accurately measured absorption lines in the 1500 nm wavelength region. This SRM can be used for high resolution applications, such as calibrating a narrowband tunable laser, or lower resolution applications, such as calibrating an optical spectrum analyzer. (Gilbert and Swann 2001, p. 2)

The main difference between the new wavelength calibration standard, SRM 2517a, and its predecessor, SRM 2517, is the use of lower pressure in the acetylene cell to produce narrower lines. Because of that difference, SRM 2517a can be used in higher resolution and higher accuracy applications.

7.2. Implications for Industry and Society

This ATP intramural project complemented the SRM-related research of the Optical Fiber and Components Group and was a natural extension of previous research related to SRM 2517. While research on SRM 2517a would have occurred in the absence of ATP's support, it would not have progressed as rapidly. According to Dr. Gilbert: "The ATP funding accelerated this project and enabled us to complete the development of a new wavelength calibration SRM about one year faster that we would have without this funding." If ATP had not funded the project, the EEEL would have invested a similar amount, but the streams of benefits and costs would have been roughly a year later.

NIST has been selling SRM 2517a at a rate of two to three per month since it was introduced in late 2000.[3]

7.3. Evaluation Approach

Wavelength references for optical fiber communications is a publicly funded, publicly performed research project, and thus the counterfactual evaluation method is appropriate. However, ATP requested that the traditional Griliches/Mansfield method be applied so that this case study could be comparable to other case studies that they funded. Thus, this case study concludes with estimates of producer and consumer surplus.

7.3.1. Benefit and Cost Information

Detailed descriptions of the uses of SRM 2517a are provided in this chapter, but in overview NIST's experience suggests that most of the test equipment manufacturers in industry use the SRM units to conduct periodic calibration checks on their equipment.[4] The calibration checks with the SRM are not typically in the production line where various intermediate standards are used for routine calibration checks. Rather, the SRM is used to check those intermediate standards. Some of these test equipment manufacturers make absorption cells—commercial versions of the SRM 2517a artifact described earlier—to incorporate into their products.

In those situations where the absorption cells are purchased, discussions with industry experts reveal that SRM 2517a is used both to check the commercial versions of the absorption cells and for study as a manufacturing guide in the production of the commercial high-volume versions of the cell. Discussions with industry show that the component manufacturers often integrate the SRM 2517a into their production lines to continuously calibrate their equipment. Network systems providers use the SRMs to calibrate their test equipment.

The industry costs and benefits for SRM 2517a were based on estimates—obtained through detailed telephone interviews over the time period from mid-June to mid-July of 2002—from industry respondents that collectively have purchased about 30% of the SRM 2517a cells.[5]

Discussions with industry identified several types of benefits and costs associated with SRM 2517a. Benefits fell within five general categories: production related engineering experimentation cost savings, calibration cost savings, yield, negotiation, and marketing. Costs were the ATP development costs plus the pull costs associated with using the SRM purchased from NIST.

Separating the SRM 2517a benefits from the benefits of other SRMs in the 25xx family was often difficult for industry respondents.[6] Some used the entire set of SRM 25xx artifacts; those respondents sometimes thought of the set of artifacts as an integrated whole, covering different parts of the spectrum of wavelengths to which equipment must be calibrated. Thus, to some extent the benefit estimates below reflect a joint benefit from the set of NIST SRM 25xx artifacts. However, there are also major sources of unmeasured industrial benefits from SRM 2517a. As a result, the benefit estimates used were, on balance, conservative for at least three reasons. First, the estimates are truncated after 10 future years, even though some respondents believed that the commercial usefulness of SRM 2517a would extend well beyond that period. Second, and more important, many respondents could not quantify the loss in sales, and therefore profits, which would occur without traceability to NIST of their wavelength calibrations. And third, the benefit estimates reflected only the benefits to the purchasers of the NIST SRM 2517a artifacts; they did not capture the additional benefits to users further down the supply chain.[7] Given these sources of downward bias, we believe that, on balance, the benefit estimates used to compute the evaluation metrics to characterize the outcomes of SRM 2517a were conservative.

Use of SRM 2517a resulted in the following:

- Production-related engineering and experimentation cost savings
- Calibration cost savings
- Increased production yields
- Negotiations cost savings
- Reduced marketing costs

7.3.2. Production Related Engineering and Experimentation Cost Savings

Users of SRM 2517a regularly conducted what we call "production-related engineering experimentation." These activities were an important aspect of production. The new more accurate measurement technology associated with SRM 2517a lowered the cost of these activities and hence represented a cost-savings benefit. Also experimentation costs for industry have been lowered because of industry's interaction with the NIST scientists who developed the artifact. One industry expert said SRM 2517a reduced his firm's investigation costs and added that, without it, the firm would have invested additional engineering person-years with equipment to maintain production.[8]

7.3.3. Calibration Cost Savings

SRM 2517a reduced the costs of calibrating production equipment and products. Examples, based on discussions with respondents in industry, include the observation of one respondent that it was not uncommon to recalibrate production devices for an optical fiber network on a daily basis, or even more frequently. The respondent said that SRM 2517a reduced the cost of each calibration; it permitted equipment to be calibrated on the production floor. The alternative, according to the respondent would be to purchase tunable lasers, which not only are more costly but also must be set for one frequency at a time, whereas the SRMs provide a fingerprint covering a whole range of the spectrum of wavelengths. In addition, tunable lasers entail additional operating time using well-trained technicians involved in production.

A respondent, whose firm manufactures locked lasers and gas cells, observed that the alternative to SRM 2517a for calibration is to invest in a suite of equipment and then take the extra time to get the calibration results. A telecommunications firm responded that prior to SRM 2517a it relied on its own internal standards based on one frequency and then extrapolated to other frequencies. The firm's expert stated that the SRM 2517a standard, with multiple indicators of various frequencies, was a critically important advance for telecommunications.

7.3.4. Increased Production Yields

One expert reported that production yields increased because SRM 2517a improved process control and thereby reduced the costs of product failure, observing: "Manufacturers of optical fiber network components produce to the customer's specifications and needs. SRM 2517a, as well as other SRMs in the 25xx series, provided useful reference points across a stable wavelength range for the tuning of the components for optical communications systems. As a costly and less accurate alternative, the points of reference could be simulated with cascades of optical filters strung together."

A manufacturer of narrow band optical filters told us that by using SRM 2517a his firm avoided a 30% loss in production yield: "SRM 2517a provided narrow line widths for reference to absolute vacuum wavelengths, and this is critical to meeting the performance specification needs of our customers. This artifact gave us an unquestionable reference to absolute wavelengths so that secondary standards can be recalibrated as they drift. Our alternative, over say 30 nm of wavelength range for a particular product, maybe 10 optical filters would be strung together. While the cost of this alternative was not that great, performance tolerances and wavelength stability would be lost. Using the alternative would have resulted in a yield loss of about 30%."

7.3.5. Negotiations Cost Savings

Negotiations with customers over disputes about the performance attributes of products were reduced because of SRM 2517a and the traceability to the NIST standard that it provides. In the absence of wavelength stability, manufacturers and customers would both have grounds to disagree about performance characteristics. Without SRM 2517a and the traceability that it provides, costly negotiations and testing would occur. One respondent said that without NIST traceability through SRM 2517a, interactions with the customers over performance characteristics would be like dealing with "a wound that would not heal."

7.3.6. Reduced Marketing Costs

Respondents reported that marketing costs were reduced because SRM 2517a allows traceability of an important new standard to NIST, and sales are greater than for SRM 2517 because of the confidence inspired by the new standard traceable to NIST. Paraphrasing a component manufacturer: There were two parts to the sales/marketing impact of SRM 2517a for our firm. First, there was a savings in personnel costs because there was less effort needed to convince customers about the quality and reliability of our products. More important, there was a positive effect on our reputation and the customers' confidence in our product line because of having NIST standards integrated in the production process. That positive effect translated into extra sales and extra profits. Paraphrasing a manufacturer of wavelength meters: We used SRM 2517a as we manufactured wavelength meters. SRM 2517a was used to check periodically the calibration of test lasers and equipment used for the qualification of our wave meters. We can claim traceability to NIST. There were cost savings to us in the sales/marketing category.

7.4. Quantitative Estimates

Quantitative estimates of each of the aforementioned categories of benefits were obtained from the five manufacturers with whom we spoke. According to Dr. Gilbert, these five firms collectively had purchased about 30% of the SRM 2517a cells sold to date. The benefit data in Table 7.1 capture industry-wide benefits. Each datum in Table 7.1 (and in the third column in Table 7.2) is the product of the sum of the dollar values for each respondent multiplied by 3.33 (3.33 = 1/0.30), and all dollar values are converted to year 2000 dollars (at the request of ATP).

To be conservative, the estimated benefits from SRM 2517a were truncated after 10 years. Respondents indicated that the SRM 2517a provided knowledge that would be commercially useful for the foreseeable future. Some respondents emphasized that, as a standard, the knowledge embodied in SRM 2517a would last and be useful virtually forever. However, industry could require even more development of the standards for measuring the wavelength of light as time passes, and the respondents as a group believed that a commercial lifetime of 10 years would be a conservative estimate for the period of intense industrial use of SRM 2517a.

Table 7.1 Industry Benefits Truncated after 10 Years (Year 2000 Dollars)

Year	Production Cost Savings ($1000s)	Calibration Cost Savings ($1000s)	Increased Production Yield ($1000s)	Decreased Negotiation Costs ($1000s)	Decreased Marketing Costs ($1000s)
1999	$3,193.9				
2000	$3,266.5	$401.0	$2,613.3	$245.0	$473.7
2001	$1,388.3	$1,832.6	$10,531.7	$1,094.3	$1,894.7
2002	$1,682.3	$353.5	$2,106.3	$218.9	$383.0
2003	$1,388.3	$441.8	$2,632.9	$273.6	$478.8
2004	$1,388.3	$589.7	$3,514.2	$365.2	$639.0
2005	$1,388.3	$735.8	$4,384.6	$455.6	$797.3
2006		$846.8	$5,046.5	$524.4	$917.7
2007		$973.8	$5,803.2	$603.0	$1,055.2
2008		$1,119.9	$6,673.7	$693.4	$1,213.5
2009		$1,287.8	$7,674.7	$797.5	$1,395.6

Note: Production-related engineering and experimentation cost savings decrease in 2001. Although some experimental production uses of the measurement technology were reported after the introduction of the SRM, the most intense realization of such experimental benefits came from the application of the new measurement technology—gained in industry's interaction with NIST through publications, presentations, and ongoing interaction with the researchers—to production problems encountered by industry as it coped with the need for the actual improved SRM and substituted experimentation for it. Publications about the SRM 2517a technology started to appear in 1999. The other categories of industry benefits increase after the introduction of the SRM 2517a because those benefits reflect the actual use of the SRMs once they were available for use.

The observed variance in benefits (in year 2000 dollars) through time reflects three key facts. First, there are different periods of primary incidence for the various cost savings. For example, production-related engineering cost savings occurred primarily in the early years of the time series and in some cases even before the introduction of SRM 2517a.[9] In contrast, the costs of reduced yields (benefits of increased yields) were avoided throughout the time series after SRM 2517a was introduced and the technology transferred to industry. Second, the introduction of SRM 2517a occurred in late 2000 and partial-year benefits are reported; benefits increased in subsequent years because the SRM was used throughout each year. Third, the variance over time reflects the collapse of optical fiber communications industry sales from record highs in 2000–2001 to low levels in 2002. Projections by industry then reflected an expected recovery of industry sales to the levels experienced in 1999—levels that in 1999 were between one-third and one-half of their subsequent peaks

Table 7.2 Estimated Costs Associated with SRM 2517a (Year 2000 Dollars)

Year	ATP Funds ($1000s)	Industry Pull Cost ($1000s)
1998	$72.6	
1999	$76.7	
2000		$16.3
2001		$73.5

ATP, Advanced Technology Program.

in 2000–2001 before the bubble burst—by 2004–2005. Thereafter, the projections reflect what knowledgeable industry observers expected to be a 15% rate of growth.

The costs associated with the SRM 2517a project are in Table 7.2. The actual costs of the ATP intramural project are shown along with estimates of the pull costs for industry. Respondents were asked to estimate any initial costs, over and above any fees paid to NIST for SRM 2517a, to be able to use (i.e., to pull in) the artifact in production. These pull costs were one-time costs. Table 7.3 aggregates the cost and benefit estimates from Tables 7.1 and 7.2.

7.5. Results of the Economic Analysis

Table 7.4 summarizes the three evaluation metrics for this case study. Based on one or all of the metrics in Table 7.4, the ATP intramural funded SRM 2517a project was successful from society's economic perspective. The public's extant investment in the Optical Fiber and Components Group enabled a relatively small incremental investment in a refined optoelectronic standard to yield a large social benefit. At ATPs request, present values were referenced to 1998. The internal rate of return is calculated to be 4,400%; using the OMB (1992) recommended 7% real discount rate, the benefit-to-cost ratio is 267-to-1 and the net present value in year 2000 dollars is $58.1 million.

The metrics in Table 7.4 reflect only a portion of the social return on investments, and these are the returns that economists call producer surplus. Producer surplus is the profit resulting because of the use of the infrastructure technology embodied in SRM 2517a. Although the estimate will be a rough one, we were also able to provide a first-order approximation of the consumer surplus gains.

Table 7.3 Estimated Total Costs and Estimated Total Industry Benefits Associated with SRM 2517a (Year 2000 Dollars)

Year	Total Costs ($1000s)	Total Industry Benefits ($1000s)
1998	$72.6	
1999	$76.7	$3,193.9
2000	$16.3	$6,999.5
2001	$73.5	$16,741.6
2002		$4,744.0
2003		$5,215.4
2004		$6,496.4
2005		$7,761.6
2006		$7,335.4
2007		$8,435.2
2008		$9,700.5
2009		$11,155.6

Figure 7.1 represents the situation for the typical firm selling a differentiated product that used SRM 2517a in the production process.[10] The availability of the new standard reference material lowers the unit costs as shown in the figure from "unit cost 2517" to "unit cost 2517a." Consequently, the firm chooses a lower price and sells more of its product or service.[11] The firm's profit-maximizing price falls from P_1 to P_2, and the optimal output increases from Q_1 to Q_2. The new surplus—resulting because of the new lower unit costs of production enabled by SRM 2517a—is the sum of the areas A, B, C, and D. Area A represents the new producer surplus on sales of the original amount of output. Area B plus area C represents the new producer surplus from the sale of additional output. Finally, area D represents the net gain in consumer surplus (new consumer surplus that does not simply offset a loss in previously existing producer surplus).

Details about price, output, and unit cost are considered highly confidential, and the industry respondents were typically unwilling to

Table 7.4 Evaluation Metrics for the SRM 2517a Case Study

Metric	Estimate
Real internal rate of return	4,400%
Benefit-to-cost ratio	267-to-1
Net present value (using 1998 as base year in year 2000 dollars)	$58.1 million

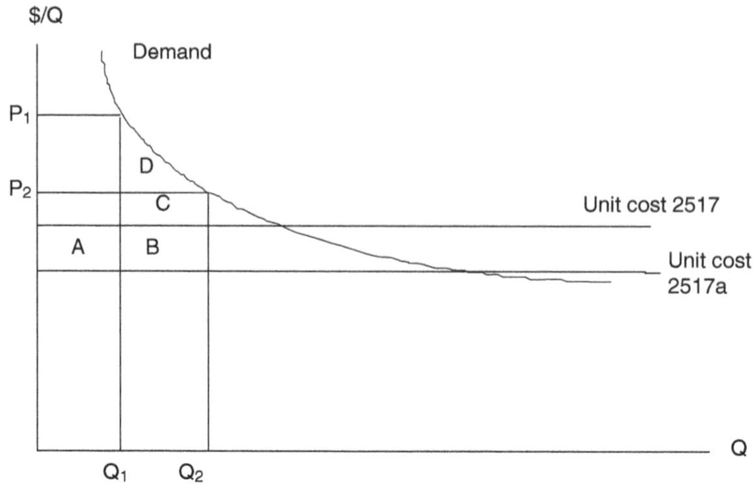

Figure 7.1 Demand, unit cost, and net gain in producer and consumer surplus from the use of SRM 2517a.

provide such information. However, one of the respondents was willing to provide detailed information, for its own production, about P_1, P_2, Q_1, Q_2, and unit cost both before SRM 2517a was introduced and then after it replaced SRM 2517. For that firm, the ratio of net new consumer surplus to new producer surplus, D/(A + B + C), equals 0.238. That firm conjectured that its experience with the cost-lowering effect of replacing SRM 2517 with SRM 2517a would be similar to the experiences of others in the industry. Therefore, as a first-order approximation of consumer surplus gains because of the process innovations from applying SRM 2517a, we multiplied the new producer profits—the industry benefits column of Table 7.3—by 0.238. Table 7.5 recalculates the metrics for the SRM 2517a project by using the total of the net gains in producer and consumer surplus (the industrial benefits from Table 7.3 multiplied by 1.238) as the social return on the investment.

Table 7.5 Revised Evaluation Metrics for the SRM 2517a Case Study Using Total Benefits (Net Gains in the Total of Producer Surplus and Consumer Surplus; Year 2000 Dollars)

Metric	Estimate
Real internal rate of return	5,500%
Benefit-to-cost ratio	331-to-1
Net present value (using 1998 as the base year in year 2000 dollars)	$72.0 million

Notes

1. This case study was conducted by Link and Scott in 2002; this case study draws from Link and Scott (2004a; 2005b) and the Link and Scott references therein.

2. Because of fundamental molecular absorptions when light is projected through the absorption cell filled with acetylene gas, the power transmitted through the cell is distinct at specified wavelengths, allowing accurate references to those wavelengths. Those references can then be used to calibrate instruments for industry.

3. The breakdown of all of the SRM 2517a sales by industry category in 2002 was approximately 45% to manufacturers of test equipment, 30% to manufacturers of components, 10% to companies providing network systems, and 15% to other users—mostly research laboratories—of the SRM. According to Dr. Gilbert, a firm will typically purchase one SRM 2517a.

4. The data developed for discussion of the outcomes in this case study were based on discussions with Dr. Gilbert and several industry experts from Wavelength References, Burleigh Instruments, Corning, Agilent, and Chorum Technologies.

5. The information about the industry-wide coverage of our sample of respondents in industry was provided by NIST.

6. For a discussion of other optoelectronics SRMs, see http://patapsco.nist.gov/srmcatalog/tables/view_table.cfm?table=207-4.htm

7. As one respondent whose firm manufactures commercial gas cells (based on SRM 2517a) for use in instruments stated: "If there were no SRM 2517a, all along the way through the supply chain the additional calibration expenses (suites of equipment and extra labor costs) would be incurred. Roughly one-half of the optical spectrum analyzers sold to industry incorporate the SRM 2517a technology to calibrate better. There would be extra expense and time at each research site."

8. See also the note to Table 7.1.

9. Industry interacts with NIST and stays abreast of the latest developments through direct communication with NIST scientists, and through scientists' presentations and publications. In this case, some respondents reported that they began benefiting from the new knowledge—gained from interaction with NIST researchers—about wavelength calibration even before SRMs were sold as industry coped with the need for the actual SRMs but substituted experimental work in their absence.

10. As is seen in Figure 7.1, in addition to gaining new profits that we have already identified as industrial benefits, industry loses some of its previous profits on the previous amount sold before unit costs fell because the use of SRM 2517a lowered costs and consequently price fell. However, those lost profits (lost producer surplus) were completely offset by a gain of exactly that amount in consumer surplus, leaving just the new profits measured in Table 7.3 and represented by (A + B + C) in Figure 7.1 as the increase in total surplus because of increased producer surplus. The net gain in consumer surplus (represented by D in Figure 7.1) is then added to get the change in total economic surplus that is the social return to the use of SRM 2517a—consumers gain more than D, but that additional gain is exactly offset by an equal amount of lost previously existing surplus for producers, leaving D as the net gain in consumer surplus.

11. Note that Figure 7.1 depicts optimal output in the long run when all costs are variable.

8

The Malcolm Baldrige National Quality Award

8.1. Background Information

In response to the productivity decline in the non-farm U.S. economy in the mid-1970s and early-1980s, a number of economic policy initiatives were introduced in the early- and mid-1980s in an effort to reverse the downward productivity trend by stimulating innovative activities within firms.[1] One such initiative was the Malcolm Baldrige National Quality Improvement Act of 1987 that declared[2]:

> [T]he leadership of the United States in product and process quality has been challenged strongly (and sometimes successfully) by foreign competition, and our Nation's productivity growth has improved less than our competitors over the last two decades; . . . a national quality award program . . . in the United States would help improve quality and productivity by—
>
> (A) helping to stimulate American companies to improve quality and productivity for the pride of recognition while obtaining a competitive edge through increased profits,
>
> (B) recognizing the achievements of those companies which improve the quality of their goods and services and providing an example to others,
>
> (C) establishing guidelines and criteria that can be used by businesses, industrial, governmental, and other organizations in evaluating their own quality improvement efforts, and

(D) providing specific guidance for other American organizations that wish to learn how to manage for high quality by making available detailed information on how winning organizations were able to change their cultures and achieve eminence.

[And] there is hereby established the Malcolm Baldrige National Quality Award.

Physically and administratively located at the National Institute of Standards and Technology (NIST), the Baldrige National Quality Program has been supported by federal and private funding since its establishment.[3] On the federal side, support for the program has increased from $200,000 in 1988 to $5,344,000 in 2000.

Although the program is small by budgetary standards, the program might be one of the better known competitiveness programs sponsored by the government. Certainly, the Malcolm Baldrige National Quality Award has been one of the most widely publicized of all public-sector-sponsored performance awards.

Although the program's federally funded budget has increased since its inception, as shown in Figure 8.1, the number of applicants to the program for the award has not, as shown in Figure 8.2. There has been speculation that the decline in the number of applicants since 1991 reflects the increasing opportunity costs for organizations to conform to the Baldrige Criteria for Performance Excellence listed in Table 8.1. The number of applicants for 1999 and 2000 increased because the program's scope was broadened to include awards for educational organizations and health care providers.

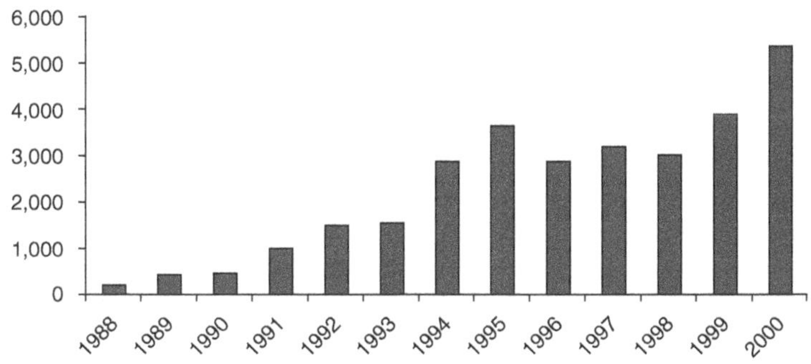

Figure 8.1 NIST allocations to the Baldrige National Quality Program ($1000s), 1988–2000.

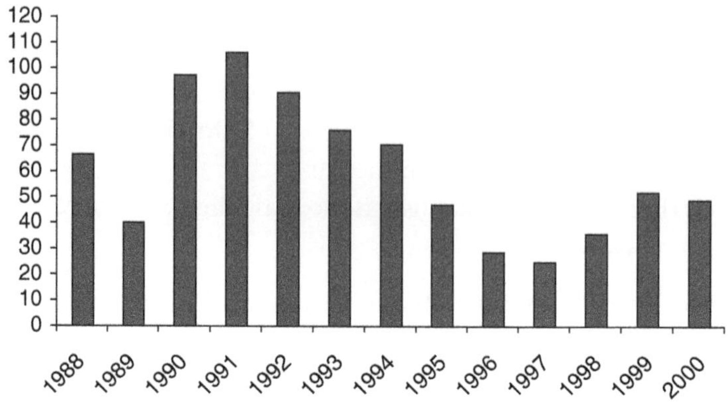

Figure 8.2 Applications to the Baldrige National Quality Program, 1988–2000.

Table 8.1 Overview of the Baldrige Criteria for Performance Excellence

The Baldrige Criteria for Performance Excellence provide organizations with an integrated, results-oriented framework for implementing and assessing processes for managing all operations. The Baldrige Criteria are also the basis for making the Baldrige Award and providing feedback to applicants. The Baldrige Criteria consist of seven categories:

- *Leadership:* The firm's leadership system, values, expectations, and public responsibilities.
- *Strategic planning:* The effectiveness of strategic and business planning and deployment of plans, with a strong focus on customer and operational performance requirements.
- *Customer and market focus:* How the firm determines customer and market requirements and expectations, enhances relationships with customers, and determines their satisfaction.
- *Information and analysis:* The effectiveness of information collection and analysis to support customer-driven performance excellence and marketplace success.
- *Human resource development and management:* The success of efforts to realize the full potential of the workforce to create a high performance organization.
- *Process management:* The effectiveness of systems and processes for assuring the quality of products and services.
- *Business results:* Performance results, trends, and comparison to competitors in key business areas—customer satisfaction, financial and marketplace, human resources, suppliers and partners, and operations.

Source: NIST (1997b).

The firms that had received the Malcolm Baldrige National Quality Award through 2000 (the year prior to this study) are listed in Table 8.2. At that time, 43 awards had been announced.

One preliminary indication of the social benefits associated with the program is, as shown in Table 8.3, that the award criteria have been adopted by states as a foundation or benchmark for their own quality award programs, thus signifying one dimension of spillover benefits.

Table 8.4 provides a concise perspective of the extant empirical economics and management literature (published at the time of this case study) related to measurable firm performance effects associated with all aspects of the program. The table shows that scholars have identified several aspects of the social benefits associated with the program, the award, and the underlying criteria. Therefore, our a priori expectations were that the net social benefits associated with the program were substantial.

8.2. Application of the Counterfactual Evaluation Method

In a broad sense, the Baldrige National Quality Program is a measurement-and-standards infrastructure research and development investment program, with the associated investments in operations and maintenance. Publicly funded, publicly performed infrastructure R&D and related operations and maintenance investments occur within the program in the sense that therein the Baldrige Criteria were originally developed and therein, through the Baldrige Award process, appropriate applications of the criteria for performance excellence are evaluated. In this broad sense, the Baldrige National Quality Program is similar to a NIST laboratory that performs infrastructure technology R&D investments and sets performance standards (i.e., the Baldrige Criteria) and then continually calibrates bench standards used in private-sector laboratories to achieve a predetermined level of performance (i.e., the Baldrige Award process).

Thus, we applied the counterfactual evaluation method to the evaluation of the Baldrige National Quality Program. Benefits to the economy from the program are systematically quantified in terms of the cost savings that organizations realized by having the Baldrige Criteria to follow as opposed to organizations, on their own, developing and testing comparable criteria.

Table 8.2 Malcolm Baldrige National Quality Award Winners, 1988–2000

Year	Recipients
1988	Motorola Inc.
	Westinghouse Electric Corporation, Commercial Nuclear Fuel Division
	Globe Metallurgical Inc.
1989	Milliken & Company
	Xerox Corporation, Business Products and Systems
1990	Cadillac Motor Car Company
	IBM Rochester
	Federal Express Corporation
	Wallace Co., Inc.
1991	Solectron Corporation
	Zytec Corporation
	Marlow Industries, Inc.
1992	AT&T Network Systems Group, Transmission Systems Business Unit
	Texas Instruments Incorporated, Defense Systems & Electronics Group
	AT&T Universal Card Services
	The Ritz-Carlton Hotel, Company
	Granite Rock Company
1993	Eastman Chemical Company
	Ames Rubber Corporation
1994	AT&T Customer Communications Services
	GTE Directories Corporation
	Wainwright Industries, Inc.
1995	Armstrong World Industries Inc., Building Products Operation
	Corning Incorporated, Telecommunications Products Division
1996	ADAC Laboratories
	Dana Commercial Credit Corporation
	Custom Research, Inc.
	Trident Precision Manufacturing, Inc.
1997	3M Dental Products Division
	Solectron Corporation
	Merrill Lynch Credit Corporation
	Xerox Business Services
1998	Boeing Airlift and Tanker Programs
	Solar Turbines, Incorporated
	Texas Nameplate Company, Inc.
1999	STMicroelectronics, Inc.-Region Americas
	BI
	The Ritz-Carlton Hotel Co., LLC
	Sunny Fresh Foods
2000	Dana Corp.-Spicer Driveshaft Division
	KARLEE Company, Inc.
	Operations Management International, Inc.
	Los Alamos National Bank

Source: http://www.quality.nist.gov

Table 8.3 Application of Baldrige Criteria to State and Local Quality Award Programs

Year	No. of States with Award Programs Tied to Baldrige Criteria	No. of Service and Manufacturing Organizations that Applied for State and Local Quality Awards
1991	8	111
1992	12	144
1993	19	357
1994	29	428
1995	37	574
1996	42	804
1997	43	974
1998	44	830

Source: Blodgett (1999) and NIST (1998).

Table 8.4 Empirical Literature Related to Measurable Firm Performance Effects Associated with the Baldrige National Quality Program

Aspects of Performance	Authors
Award winners have stronger financial performance and greater increases in their market value.	Wisner and Eakins (1994) Helton (1995) Lawler, Mohrman, and Ledford (1995) Huselid and Becker (1996) Easton and Jarrell (1998) NIST (1996, 1997a, 2000) Hendricks and Singhal (2001)
Application of Baldrige Criteria improved employee relations, lowered costs, and improved customer satisfaction.	GAO (1991)
Application of Baldrige Criteria improved competitiveness.	Council on Competitiveness (1995) Banker, Khosla, and Sinha (1998)
Application of Baldrige Criteria increased worker productivity.	Ichniowski, Shaw, and Prennushi (1995) Black and Lynch (1996a, 1996b)

Note: This table summarizes the extant literature at the time (2001) the original study, on which this chapter is based, was completed.

As discussed in the following section, benefit data were collected through surveys of selected members of the American Society for Quality, and then extrapolated cautiously to the aggregate economy. Cost data were provided by the Baldrige National Quality Program Office at NIST. The evaluation metric requested by the Program Office was a benefit-to-cost ratio, with all benefits and all costs referenced to year 2000.

The American Society for Quality (ASQ),[4] formed in 1946, agreed to a request from the management of the Baldrige National Quality Program Office to have a mail survey distributed to its 875 U.S. private-sector firms and public-sector organizations (hereafter, members).[5]

ASQ's stated mission is to advance individual and organizational performance excellence on a worldwide basis by providing members opportunities for learning, quality improvement, and knowledge exchange. As stated on its Web site, the Society's objectives for 2000 were as follows:

- To be our members' best resource for achieving professional and organizational excellence
- To be a worldwide provider of information and learning opportunities related to quality
- To be the leader in operational excellence and delivering customer value
- To be the recognized leader worldwide for advancing individual and organizational performance excellence

8.2.1. Social Costs of Operating the Baldrige National Quality Program

The public source of funds for the Baldrige National Quality Program is an annual allocation from the NIST budget. Column (2) in Table 8.5 shows the program's annual allocations from NIST by fiscal year beginning with its first year of operation, 1988, and ending with 2000.

The Malcolm Baldrige National Quality Improvement Act of 1987 stated that: "The Secretary [of Commerce] is authorized to seek and accept gifts from public and private sources to carry out the program."

In addition to the public funding through NIST, there have been private sources of funds. The program was initially endowed by private industry with $10 million. A foundation was established to manage these funds and to allocate the interest earned to the program for award ceremonies, publication costs, and partial training and travel costs for examiners

Table 8.5 Baldrige National Quality Program Operating Costs

(1) Fiscal Year	(2) NIST Allocations ($)	(3) Foundation Allocations ($)	(4) Firm Reimbursed Examiner Expenses ($)	(5) Examiner Time (hours)	(6) Total Operating Costs (year 2000 dollars)
1988	$200,000	*$600,000*	*$190,000*	*37,995*	$3,689,349
1989	408,000	*600,000*	*190,000*	*37,995*	3,910,205
1990	488,000	*600,000*	*190,000*	37,995	3,951,030
1991	1,018,000	*600,000*	*190,000*	46,510	5,059,093
1992	1,482,000	*600,000*	*190,000*	49,763	5,750,259
1993	1,525,000	*600,000*	*190,000*	46,223	5,516,050
1994	2,860,000	728,973	190,453	45,944	7,072,918
1995	3,611,000	694,669	188,137	51,259	8,092,820
1996	2,865,000	652,017	160,230	44,143	6,683,663
1997	3,174,000	778,600	171,803	44,090	7,073,404
1998	3,010,000	808,713	157,879	43,662	6,840,293
1999	3,877,000	1,159,337	186,052	51,735	8,553,566
2000	5,334,000	1,187,543	160,363	51,349	9,891,218

Notes: Column (2): NIST allocation data were provided by the Award office. For inclusion in column (6) these data were inflated to year 2000 dollars using the chain type price index for gross domestic product (GDP) from Table B-7, "Chain type price indexes for gross domestic product, 1959–2000" Council of Economic Advisers (2001, p. 284).
Column (3): Foundation allocation data were provided by the Award office for 1994–2000. The upper bound on pre-1994 data was estimated (*italics*), with advice from the Award office. For inclusion in column (6) these data were inflated to year 2000 dollars using the chain-type price index for GDP in Council of Economic Advisers (2001, p. 284).
Column (4): Foundation reimbursements of 70% were paid in 1999 and 2000 for examiners in the education and health care areas; all other examiners were reimbursed at 60% of their expenses. From these data, provided by the Award office, firm reimbursed expenses were calculated for 1994–2000. The upper bound on pre-1994 firm cost was estimated (*italics*), with advice from the Award office. For inclusion in column (6) these data were inflated to year 2000 dollars using the chain-type price index for GDP in Council of Economic Advisers (2001).
Column (5): Examiner time was provided by the Award office. The upper bound on pre-1990 examiner time was estimated (*italics*), with advice from the Award office. Based on the management background of the numerous examiners involved in the program, the Award office estimates that the current fully burdened value of a year of examiner time is $125,000 (year 2000 dollars based on 2,000 hours per year). The estimated value of examiner time is included in column (6) without additional adjustment.
NIST, National Institute of Standards and Technology.

whose firms would not pay for such expenses. In column (3) of Table 8.5 are the program's annual allocations from the foundation. In column (4) are annual estimates of firm and organization expenditures for examiner travel that were not reimbursed by the foundation through the program.[6]

Industry also supports the program through volunteer examiners during the application and evaluation process. In column (5) of Table 8.5 are the total hours of examiner time that were devoted to training, application review, and site visits.

Column (6) of Table 8.5 reports the estimated program costs (in year 2000 dollars), by year. The present value of these costs, brought forward to year 2000 at the request of the Baldrige National Quality Office (rather than referenced to 1988) at the OMB (1992) recommended real rate of return of 7%, is $118,617,000.

Thus, $119 million (rounded in year 2000 dollars) was used to represent the present value of the total social costs through 2000 associated with the Baldrige National Quality Program.

8.2.2. Social Benefits Associated with the Baldrige National Quality Program

A five-step approach was used to estimate the net social benefits associated with the Baldrige National Quality Program.

Benefit data were collected by survey from a sample of the membership of ASQ. These benefit data were extrapolated first to the ASQ membership as a whole and then cautiously to the economy as a whole. The present value of the conservative estimate of the net private benefits received by the ASQ members as a result of the Baldrige National Quality Program was $2.17 billion (rounded in year 2000 dollars).

If the entire economy benefited to the same extent as the ASQ members, the present value of the conservative estimate of the net social benefits associated with the Baldrige National Quality Program was $24.65 billion (rounded in year 2000 dollars).

The net private benefits to ASQ members and net social benefits were estimated as follows:

Step 1: Estimating the probability of survey response from ASQ members. As noted earlier, the ASQ agreed to a request from the management of the Baldrige National Quality Program Office at NIST to distribute a survey administered by the Program Office to its 875 U.S. members. Sixty-five organizations returned completed or partially completed survey instruments.[7]

Step 1 quantified the probability that an ASQ member who received a survey would respond to the survey. Obviously, the average probability of response is 65 returned surveys out of 875 sent surveys, or a 7.43% response rate. However, for the statistical analysis that follows, an estimated probability of response for each of the 875 members is needed as a control variable used in Step 2.

The probability of a member responding to the survey was estimated using an industry effects model represented as:

$$\text{Probability of response} = \text{F (two-digit SIC industry variablles)} \quad (8.1)$$

where the dependent variable used to estimate equation (8.1) equals 1 if the member returned a completed or partially completed survey, and 0 otherwise; and where the two-digit Standard Industrial Classification (SIC) industry variable categories are as described in the notes to Table 8.6.[8]

Equation (8.1) posits that the probability of a member responding can be predicted based on the industry in which that member produces. The probit results from equation (8.1) are in Table 8.6. For each of the 875 surveyed members, the estimated model in Table 8.6 produces a predicted value for the probit index, z, for the probability of response.[9,10]

Step 2: Estimating the probability of self-assessment for responding members. Step 2 quantified the probability that an ASQ member who received a survey conducted a quality-based self-assessment. A probability of self-assessment was needed in the estimation of net benefits. First, a probability of self-assessment model is estimated and, second, a prediction of the probability of self-assessment for each ASQ member's organization was calculated in Step 3.

The probability of a member having conducted a self-assessment in the past, given that the member returned a completed or partially completed survey, is estimated using a model written as:

$$\begin{aligned}\text{Probability of self-assessment} = \text{F (two-digit SIC industry variables,} \\ \text{competitiveness variables, control variables)}\end{aligned} \quad (8.2)$$

where the dependent variable used to estimate the model equals 1 if the member responded in the affirmative to at least one of the following survey statements, and 0 otherwise:

Has your organization performed a self-assessment using the Baldrige Criteria for Performance Excellence or related criteria

Table 8.6 Probit Results for Probability of Response to the Survey (Asymptotic t-Statistics in Parentheses; n = 859)

Variable	Estimated Coefficient
dnonmin	0.743 (1.46)
dchempet	–0.008 (–0.03)
dmcneqin	–0.076 (–0.37)
dtrcomut	0.020 (0.06)
dwholret	0.035 (0.11)
dfire	–0.047 (–0.12)
dserv	–0.586** (–2.02)
dbusser	0.350 (1.49)
dhealth	0.795** (2.07)
dpubadm	–0.215 (–0.75)
constant	–1.418*** (–8.73)
log likelihood	–220.297
pseudo R^2	0.043
$Chi^2(10)$	19.94**

Notes: ***significant at 0.01 level, **significant at 0.05 level, *significant at 0.10 level.
The 16 observations in the miscellaneous category (members who could not be assigned to a two-digit SIC industry or who were assigned to miscellaneous manufacturing) were dropped because the miscellaneous category predicted nonresponse perfectly.
dnonmin = 1 for the agriculture, forestry, fisheries, minerals, and construction industries, and 0 otherwise; includes SICs < 20.
dchempet = 1 for chemicals, petroleum, and rubber, and miscellaneous plastics, and 0 otherwise; includes SICs 28, 29, and 30.
dmcneqin =1 for machinery and equipment, both non-electric and electric and electronic, and instruments, and 0 otherwise; includes SICs 35, 36, 37, and 38.
dmats = 1 for the remaining manufacturing SICs, and 0 otherwise; includes SICs 20 through 27 and SICs 31 through 34; observations with *dmats* = 1 are in the intercept.
dtrcomut = 1 for transportation, communications, and utilities, and 0 otherwise; includes all two-digit SICs greater than 39 and less than 50.
dwholret = 1 for wholesaling and retailing, and 0 otherwise; includes all two-digit SICs greater than 49 and less than 60.
dfire = 1 for finance, insurance, and real estate, and 0 otherwise; includes all two-digit SICs greater than 59 and less than 70.
dserv = 1 for other services other than business services and health services, and 0 otherwise; includes all two-digit SICs greater than 69 and less than 90 except for SIC 73 and SIC 80.
dbusser = 1 for business services, and 0 otherwise; includes SIC 73.
dhealth = 1 for health services, and 0 otherwise; includes SIC 80.
dpubadm = 1 for public administration, and 0 otherwise; includes two-digit SICs greater than 89 and less than 100.

(and by related criteria we mean criteria informed or derived by the Baldrige Criteria)? If "yes," in what year(s)?[11]

Has your organization applied for the Malcolm Baldrige National Quality Award? If "yes," in what year(s)?

Has your organization applied for a state quality award? If "yes," in what year(s)?

and where the competitiveness variables noted in equation (8.2) are defined in terms of a member's Likert responses (7 = strongly agree to 1 = strongly disagree) with the following two survey statements:[12]

The possibility or threat of new competition is significant. (*comp*)
Our customers have a significant ability to bargain on the price of our primary products. (*barg*)

and where the relevant control variables are based on estimates of the probability of response (*probres*) to the survey from equation (8.1).

The probit results from equation (8.2) are in Table 8.7.[13,14]

Step 3: Predicting the probability of self-assessment for members of ASQ. The statistical output from this Step 3 yielded an estimate of the probability of conducting a self-assessment for each of the 875 members of ASQ using the results (n = 60) from the estimated probit model presented in Table 8.7.

With reference to equation (8.2), a probit index for each of the 875 members was estimated by multiplying the actual value of each independent variable for each member by the estimated probit coefficient reported in Table 8.7.[15,16]

Step 4: Estimating the net social value of the Baldrige National Quality Program to ASQ members. Of the 23 members of ASQ that performed a self-assessment, 14 responded to the following counterfactual survey statement:

> In the absence of the Malcolm Baldrige National Quality Award—and therefore without the information and assistance that it provides about performance management/quality improvement assessments and therefore with the need to incur expenditures to develop and acquire such knowledge and assistance from other sources—what expenditures (fully burdened) would your organization have incurred to achieve the same level of expertise in performance management or quality improvement that you now have? $_____ per year over the previous _____ years.

Table 8.7 Probit Results for Probability of Self-Assessment (Asymptotic t-Statistics in Parentheses; n = 60)

Variable	Estimated Coefficient
dwholret	0.899 (1.33)
dpubadm	1.932*** (2.46)
comp	–0.189 (–1.36)
barg	0.234* (1.80)
probres	4.248 (1.25)
constant	–1.276 (–1.40)
log likelihood	–32.096
pseudo R^2	0.124
$Chi^2(5)$	9.11*

Notes: ***significant at 0.01 level, **significant at 0.05 level, *significant at 0.10 level. There are 65 observations available to estimate the model in equation (8.2); however, the two-digit industry variables, *dtrcomut* and *dfire*, are dropped along with the five observations where they equal 1 because they predict assessment perfectly. Thus, the results above are based on 60 observations.

As discussed earlier with reference to the counterfactual evaluation method, members' responses to this statement represented credible time estimates of the benefits (i.e., the costs avoided reported in year 2000 dollars) associated with the Baldrige National Quality Program. Thus, for each of the 14 responding members, a time series of real benefits received was formulated.

Regarding costs to compare to this time series of benefits, each of the 14 members responded to the following two questions:

> If your organization has been an award applicant, what was the total economic cost (fully burdened) to your organization to obtain, understand, collect relevant information, and comply with the Baldrige Criteria or state application requirements? $_____ per year during the year(s) _____.
>
> If your organization did not apply for the Malcolm Baldrige National Quality Award or state award, but nonetheless performed a self-assessment using the Baldrige Criteria or related criteria, what was the total economic cost (fully burdened) to your organization to perform the self-assessment? $_____ per year during the year(s) _____.

Thus, for each of the 14 responding members, a time series of real (in year 2000 dollars) pull costs incurred to make the Baldrige Criteria operational is also developed.

The net present value of each member's benefits was calculated using these survey data by first calculating the present value (referenced to the earlier of the first year of benefits or the first year of costs, hereafter the base year) of each member's benefits and each member's costs. The discount rate for this calculation is $r = (k - 0.03)/(1 + 0.03)$, where k is each member's reported hurdle rate (see Fig. 2.1) and where the prevailing rate of price inflation over the reported time intervals was estimated at 3%.[17] Thus, net present value is the difference between the present value of benefits less the present value of costs, both referenced to the base year. Each member's net present value of benefits was then re-referenced to 2000 using a 7% growth rate to account for the social opportunity costs of these moneys (OMB 1992).

The following model is estimated using the 14 calculated net present values:[18]

$$NPV_{2000} = \text{F(two-digit SIC industry variables, size variables)} \quad (8.3)$$

where member size was provided by ASQ for 874 of the 875 members. The least-squares results from equation (8.3) are in Table 8.8.[19]

The estimated coefficients in equation (8.3) were used to forecast the net present value of benefits for each of the 874 members of ASQ for which member size was available.

The predicted values from estimating equation (8.2) represent point estimates for the probability of each member of ASQ conducting a self-assessment. The predicted values from equation (8.3) represent point estimates of the net present value of benefits associated with the Baldrige Program conditional on a member conducting a self-assessment.

Table 8.8 Least-Squares Results for Net Present Value of Benefits (t-Statistics in Parentheses; n = 14)

Variable	Estimated Coefficient
size	–83844.49** (–2.48)
size^2	13.33** (2.27)
dtrcomut	4.90×10^7* (2.10)
constant	9.45×10^7** (2.71)
$F_{(3,10)}$	3.51*
R^2	0.513

Notes: ***significant at 0.01 level, **significant at 0.05 level, *significant at 0.10 level. The explanatory member-size variable was measured in millions of dollars, while the dependent variable for value is measured in dollars.

The product of these two estimates gives a point estimate of the expected net present value from the Baldrige Program for a member of ASQ. By using the standard errors of our predictions from the estimations of equations (8.2) and (8.3), we control, to the extent possible, for the relatively small sample of members that provided the detailed information about their net benefits from the program.

Thus, in an effort to present conservative estimates of the net present value of benefits associated with the Baldrige Program to members of ASQ, the following adjustments are made.

First, regarding the predicted values of the probability of a self-assessment from the estimation of equation (8.2), a 0.4142 confidence interval is calculated for each member of ASQ, and the lower bound on that interval is used as the relevant predicted value of the probability of self-assessment for that member. The lower bound on a 0.4142 confidence interval implies that there is a 0.7071 probability that the true value of the probability of self-assessment is greater than the value being used.[20]

Second, regarding the predicted value, conditional on self-assessment, of the net present value of benefits associated with the Baldrige Program from the estimation of equation (8.3), a 0.4142 confidence interval is calculated for each ASQ member using the standard errors for the linear combination of the estimated coefficients and for the error in equation. The lower bound on that interval is then used as the conservative net present value conditional on self-assessment by the member.

The product of the lower bound of the probability of self-assessment from the estimation of equation (8.2) and the lower bound of the net present value of benefits from the estimation of equation (8.3) yields for each member an estimate of net present value of benefits. That estimate could be lower or higher than the true value of the net present value of benefits. The true value has greater than a 50% probability ($0.7071 \times 0.7071 = 0.50$) of being larger than the value being used as the estimate, because the probability that both estimates multiplied are exceeded by their true values is 0.50. Of course, in some cases where the true value of one but not the other of the two estimates being multiplied falls short of the lower bound, the true value of net present value benefits may still exceed the estimate used. Hence, the true value has more than a 50% probability of being greater than the one used.

The sum of the conservative, lower-bound derived value of net benefits for ASQ members is \$2.17 billion.[21]

Thus, if it is assumed that there is no value associated with the Baldrige National Quality Program other than that received by the ASQ members, the conservative present value for net private benefits is $2.17 billion. When compared to the present value of the total social cost associated with the program of $119 million, the ratio of ASQ benefits to social costs is 18.2-to-1.

Step 5: Estimating the aggregate net social value of the Baldrige National Quality Program. If the entire economy benefits from the Baldrige National Quality Program to the same extent as the ASQ members,[22] then total social benefits can be forecast using the following formula:

Economy value = (value for ASQ)/(proportion taken by the ASSQ members in the 50 represented industrial sectors) (8.4)

where the proportion in the denominator is calculated to be 0.0880285.[23]

Thus, under this assumption, the conservative present value of social benefits is $24.65 billion.[24] When compared to the present value of the total social cost associated with the program of $119 million, the ratio of economy-wide benefits to social costs is 207-to-1.[25]

8.3. Ratio of Net Social Benefits to Social Costs

As derived in the previous section, the conservative estimate of the present value of aggregate economy-wide net social benefits associated with the program through 2000 is $24.65 billion (rounded in year 2000 dollars). As also explained earlier, the present value of the social costs to operate the program through 2000 is $119 million (rounded in year 2000 dollars). From an evaluation perspective, these values yield a benefit-to-cost of 207-to-1.

Extrapolating cautiously from the ASQ membership to the entire economy—under the assumption that the entire economy benefits from the program to the same extent as ASQ members—implied a social benefit-to-cost ratio of 207-to-1. Yet the organizations outside of the ASQ could benefit even more than the ASQ membership. The ASQ members represent by member size 8.8% of the 50 industrial sectors with ASQ members. Through 2000, 11 of the 43 Baldrige Awards were received by ASQ members. Further, based on the requests for Baldrige application materials and criteria as well as the many winners from outside the ASQ, many companies outside of the ASQ were using and benefiting—conceivably even more than ASQ members—from the Baldrige Criteria.

Thus, extrapolating from the net social benefits of the program for ASQ members to the economy as a whole may underestimate the true social benefits associated with the program. In that case, the social benefit-to-cost ratio derived in this case study would understate, even beyond the conservative estimation procedure used, the true benefits of the program.

Certainly the conservatively estimated benefit-to-cost ratio of 207-to-1 supports the hypothesis that the public's investment in quality-standards infrastructure was, and likely still is, worthwhile. The Baldrige National Quality Program at NIST provides another NIST standards-infrastructure investment, although in contrast to NIST's investments in infratechnologies focused on engineering, measurement, and science, the Baldrige Program is focused on management. These public investments in management standards appear to be worthwhile when evaluated using the benefit-to-cost ratio. However, while a benefit-to-cost ratio greater than 1.0 logically implies that the program is worthwhile given the standard assumptions behind evaluation analysis, the assumptions are important.

The benefit-to-cost ratio assumes that we really do know the opportunity costs of the resources invested by the public in a program.[26] Although the social benefits greatly outweigh the costs as measured, there may be even higher yields on other potential uses of the public's funds. We have simply assumed, although it was following OMB (1992), that a yield of 7% covers the opportunity costs for the public's invested funds. Further, the value of resources has been measured by the preferences given the current distribution of income. Finally, the public may want to use its investments to promote goals such as diversity that are not measured by values of resources as determined by demands for goods and services and the costs of technologies available for providing them.

The economics and management literature referenced in Table 8.4 describes benefits from the Baldrige Program because the program leverages private sector investments in quality management. Thus, the literature about the program and the award suggests that the large benefit-to-cost ratio for the Baldrige Program is realistic and not unexpected. Taken together, the literature about the program and the benefit-to-cost ratio estimated support the hypothesis that NIST's National Quality Award Program is an efficient infrastructure investment in standards that are important for the effective operation of organizations.

Notes

1. For a review of these policies, see Link (2006).

2. As Townsend and Gebhardt (1996, p. 6) explained, the origins of the Baldrige Award grew from "alarm over the Japanese challenge to the American economy," and they concluded (p. 13): "[T]he Baldrige will retain its position of importance, a position earned by being perhaps the major factor in positioning American business for the 21st century. The Baldrige didn't just shift the paradigm for American business—it defined a whole new way to go about doing things. As a result, business communities throughout the world once again can look to America to learn how to get things done."

3. This case study was conducted by Link and Scott in 2001; this chapter draws directly from Link and Scott (2001b; 2006) and the Link and Scott references therein.

4. See http://www.asq.org

5. There were (at the time of the study in 2001) in addition to these U.S. organizational members, over 200 international organizational members plus over 120,000 individual members.

6. The foundation reimburses between 60% and 70% of examiner travel costs, and the remainder is paid by the examiner's firm or organization.

7. ASQ sent an electronic reminder to each survey recipient (i.e., the individual identified in the organization) approximately 3 weeks after the initial mailing. No member-specific information is reported herein to ensure confidentiality.

8. ASQ provided the two-digit industry for approximately 75% of its members. Public domain information was used to determine the remaining classifications, including the *Thomas Register* and other Internet search mechanisms. The simple industry effects model is significant; more elaborate models that add other available characteristics of the members have no greater explanatory power—the additional variables were not statistically significant.

9. There are 16 cases that were assigned to a miscellaneous category because either a member could not be matched uniquely to a two-digit SIC industry or was assigned to miscellaneous manufacturing. None of those 16 members responded. Consequently, the categorical variable for the group predicted nonresponse perfectly, and the 16 observations were dropped from the sample used to estimate the model and assigned a probability of response of zero.

10. Based on the probit model estimated in Table 8.6, the hazard rate is also computed as $h(z) = F'(z)/(1-F(z))$, where $F(z)$ is the probability of response given the probit index z (hence, it is the cumulative density function for the standard normal variable at the value z) and $F'(z)$ is the density of the standard normal variable at z for each observation. The hazard rate is the conditional probability of response for a small increase in z. Conditional on no response for the observation, the probability of response for a small increment in z is $F'(z)dz/(1 - F(z))$.

11. Twenty-three of 65 members had performed a self-assessment.

12. The mean value of *comp* ($n = 65$) = 5.6. The mean value of *barg* ($n = 65$) = 4.6. The inclusion of these competitiveness variables follows from the economic and management literatures related to quality shown in Table 8.4. Firms facing greater competitive pressures or buyers with greater bargaining strength are expected to be more likely to invest heavily in quality management. See, for example, Lau (1996) who developed information about his responding firms' competitive environments, including for example, the possibility or threat of new competition.

13. When the hazard rate is included in equation (8.2) in place of the probability of response, the estimated probit model performed almost identically to the model reported in Table 8.7. Furthermore, other available, potential explanatory variables were insignificant and did not add importantly to the model's explanatory power.

14. The model in equation (8.2) was estimated with 65 observations, however the two-digit industry variables, *dtrcomut* and *dfire*, are dropped along with the five observations where they equal 1 because they predict assessment perfectly. Thus, the results in Table 8.7 are based on 60 observations.

15. As noted with reference to the estimation of equation (8.2), data are available for 65 members on *comp* and *barg*. The mean value of these two variables (n = 65) is imputed to the other 810 (875–65) ASQ members for predicting the probability of self-assessment.

16. The mean value of the probit index (n = 810) = –0.7041409, corresponding to a probability of assessment = 0.2602325. In the calculations that follow, a lower-bound probit index is used rather than the predicted value averaged here. Note from the foregoing footnote that there are 65 ASQ members who responded to the survey. Also there are by happenstance 65 of 875 members where *dtrcomut* and *dfire* equal 1, so there is no probit index for them from the estimation of equation (8.3)—recall from an earlier endnote that those two categories are perfect predictors of assessment—and hence n = 810. In the calculations that follow, rather than imposing a probability of self-assessment of 1.0 on each of the additional 65 members in the perfect prediction categories, the average lower-bound probability of self-assessment from equation (8.2) was imputed to them, thus producing in these instances a more conservative estimate. The average lower-bound probability, as contrasted with the average probability, is explained later in this chapter.

17. Regarding the hurdle rate, each member was asked to respond to the following statement: What is your company's hurdle rate for investments (the minimum rate of return that your company must anticipate if it is to consider new investment worthwhile)? ___ percent.

The real rate of return will be $r = (k - a) / (1 + a)$, where a is the anticipated rate of inflation. If one invests X and receives Y, the nominal return for the period is k such that $X(1 + k) = Y$ and $k = (Y - X) / X$. Given an anticipated rate of inflation a, the real rate of return r is such that $X(1 + a)(1 + r) = Y$ since that yields the rate of return r in constant dollars: $X(1 + r) = Y / (1 + a)$. Since $X(1 + a + r + ra) = X(1 + k)$, then $k = (a + r + ra)$ and $r = (k - a) / (1 + a)$.

The mean value of $k = 0.1821$.

18. The mean value of NPV_{2000} (n = 14) = $17.7 million.

19. Other available, potential explanatory variables, including various hazard rates or associated probabilities and other sector effects, were insignificant and did not add importantly to the explanatory power of the model.

20. Each tail in a 0.4142 confidence interval contains 0.2929 of the distribution, so there is 0.7071 probability (0.4142 + 0.2929) that the true value is greater than the value being used.

21. The mean value of the conservative estimate of value (n = 874) = $2,478,039.

22. This extrapolation is similar in procedure to that used by Scherer (1982).

23. The size data for industrial sectors were assembled using information in U.S. Census Bureau (1997) and Council of Economic Advisers (2001). Size data for 1997 were inflated using the chain type price index for gross domestic product from Table B-7, "Chain type price indexes for gross domestic product, 1959–2000" Council of Economic Advisers, (2001, p. 284) to be comparable with the ASQ 1999 sales data. When 1997 sector size data were unavailable, 1992 data were used and then inflated to 1999.

24. \$2.17 billion/0.088025 = \$24.65 billion.

25. All but a few ASQ members could be separated into the manufacturing sector and the service sector. Recalculating, using only these two broad industrial categories and omitting industrial categories where there are very few members (SIC < 20 sectors with only eight ASQ members) yields a conservative estimate of the aggregate manufacturing sector's net benefits of \$7.6 billion and a conservative estimate of the aggregate service sector's net benefits of \$13.0 billion. Thus, when the sum of these estimates is compared to total social costs of \$119 million, the resulting benefit-to-cost ratio is 173-to-1.

26. Scott (2000, 2009b) discusses the implications for program evaluation of the assumption that we know the opportunity costs of the scarce resources used by the program being evaluated.

9

Technologies for the Integration of Manufacturing Applications

9.1. Background Information

The Advanced Technology Program (ATP) received its first appropriation from Congress in fiscal year 1990 and held its first general competition in that same year.[1] Since 1994, ATP has sponsored a number of focused program competitions in addition to its general competitions. According to ATP:[2] "[Focused programs are] multi-year efforts aimed at specific, well-defined technology and business goals. These programs, which involve the parallel development of a suite of interlocking R&D projects, tackle major technology problems with high payoff potential which cannot be solved by an occasional project coming through the general competition."

One such focused program is the Technologies for the Integration of Manufacturing Applications (TIMA) Program. There was a TIMA competition in 1995 from which four research projects were selected to receive ATP support, and in 1997 there was a second TIMA competition from which six research projects were funded.

The overall goal of the TIMA focused program was to develop and demonstrate the technologies needed to create affordable, integrable manufacturing systems. Many manufacturing firms need to respond more rapidly to changing markets and evolving opportunities if they are to remain competitive in global markets. Although this need is widely recognized, manufacturers find it difficult to implement the technologies needed for them to become more agile producers. Even highly automated plants and factories struggle to adapt successfully and efficiently and reconfigure production processes to accommodate design changes and new product lines. Customized systems integration efforts are often needed to achieve such changes, but they are not undertaken primarily because of idiosyncrasies in manufacturing software and incompatibilities among software applications.

Typically, factory-floor information systems focus on the operation of production equipment and the control of processes. The systems communicate neither directly nor regularly with administrative information systems, or with design and engineering systems. As a result, upstream information systems are unaware of important manufacturing details. Middle-level information systems, known as manufacturing execution systems (MESs), bridge this critical information gap.

MES solutions, complex and burdensome as they may be, can be solved by contracting with a large systems provider or integrator. However, once a manufacturer has incurred such a substantial investment it is likely to become dependent on that single vendor and thereby become unaware of, or if aware likely ignore, other vendors that may have more economical or innovative solutions. Because the initial solution involves a re-engineering of the manufacturer's business processes to be compatible with the vendor's requirements, even large manufacturers that can afford the up-front investment cost will bypass the use of MES technology.

TIMA technologies were expected to benefit a range of firms: firms that employ MES by providing them with a wider range of powerful, integrable applications that would dramatically improve the manufacturer's ability to reconfigure, scale, and adapt their processes; small- and medium-sized manufacturers by making MES more affordable and by providing a direct path toward greater automation through incremental addition of compatible applications; and vendors of MES products by expanding the market, lowering barriers to entry, and stimulating innovation and technical specialization. Consumers could benefit from the adoption of these technologies in at least two ways: a higher quality product and a lower priced product to the extent that greater automation available to a wider range of firms lowers costs and increases competition.

ATP identified a contact person for each of the 10 ATP-funded TIMA projects.[3] We interviewed seven individuals corresponding to 8 of the 10 funded projects.[4] Each of these individuals was asked the following question:

> In the absence of ATP funding, would this research have been undertaken?

Three of the eight projects are single participant projects and five are joint ventures. All five of the research joint-venture respondents answered that no, in the absence of ATP funding, the joint venture would not have been formed and that in the absence of the joint venture their firms would not have undertaken the research. For the three single participant projects, one respondent reported that the research would not have been undertaken without ATP funding. One reported that maybe the research would have been undertaken, but it would have been at a reduced level. One reported that maybe the research would have been undertaken, but it would have been at a slower pace.

Respondents were asked to describe their understanding of the characteristics of technical risk (the project being technically successful) and market risk (the project being commercially successful) associated with the research project, and how that risk affected the fact that their firm could not have undertaken the research at all or at the same level, scope, and speed in the absence of ATP funding. Of course, risks inherent in the project are viewed in the context of the respondent's firm, and we cannot be certain to what degree the perceived risk reflects risk inherent in the project and to what degree it reflects the capabilities of the firm. Table 9.1 was used as a summary device for generalizing from the responses given during the interviews and tallying the responses indicating the various factors causing barriers to investment in TIMA technologies. Again, as discussed in Chapter 2, these barriers to technology are listed as if they are discrete phenomenon, when in reality they are interrelated.

Table 9.1 Factors Creating Barriers to Technology in the Technologies for the Integration of Manufacturing Applications (TIMA) Projects

Factor	Count
1. High technical risk associated with the underlying R&D	2
2. High capital costs to undertake the underlying R&D with high market risk	6
3. Long time to complete the R&D and commercialize the resulting technology	1
4. Underlying R&D spills over to multiple markets and is not appropriable	7
5. Market success of the technology depends on technologies in different industries	2
6. Property rights cannot be assigned to the underlying R&D	5
7. Resulting technology must be compatible and interoperable with other technologies	7
8. High risk of opportunistic behavior when sharing information about the technology	0

More than eight responses are recorded in Table 9.1; there were numerous occasions where we inferred from the respondent's discussion that there was more than one barrier to technology, that is, more than one reason for their underinvestment in R&D.

The tally in Table 9.1 captures our interpretation of the discussions with the project leaders of the TIMA projects. As such, the tally might underestimate the magnitude of the importance of some of the barriers to technology that can lead to market failure. Given this caveat, at least two interesting patterns emerged from Table 9.1. One, the TIMA research projects were characterized by both technical and market risk. Technical risk is explicitly cited only twice, although it might be evident in the cases where there were concerns about the large capital cost needed to undertake the research and in the cases of concern about interoperability with other technologies. Market risk was primarily evident in the need for the resulting technology (generic technology to be applied to software development) to interface with users' information technology. The respondents discussed with us each aspect of risk, and in a second pattern of their responses, they emphasized the interfacing issues and interoperability issues. We concluded those issues were the major sources of risk contributing to market failure.

9.2. Toward a Public/Private Partnership Competition Policy

Jaffe (1998, p. 18) argued that for ATP to be effective in achieving its statutory objectives, it "must try to determine which projects proposed to it will generate large spillovers. . ."[5] His arguments that led to this conclusion assumed that ATP should only select those projects that would not be funded by the private sector in the absence of ATP funding, or if funded would be funded at a considerably lower level so that only partial results would have been realized and these results would have taken longer to occur. Jaffe pointed to a number of ATP-sponsored studies and an independent study by the General Accounting Office (GAO). These studies concluded that ATP grantees are of the opinion that their research would not have taken place in the absence of ATP funding or would have been funded at a considerably lower level and hence would have taken place only partially and would have taken considerably longer to complete. Figure 9.1, based on Figure 2.1 in Chapter 2, illustrates Jaffe's conclusion.[6]

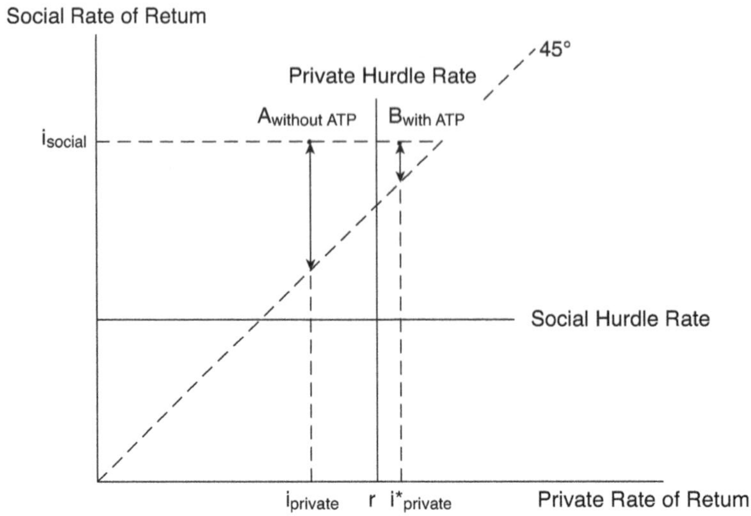

Figure 9.1 Spillover gap between social and private rates of return to Advanced Technology Program (ATP)-funded Technologies for the Integration of Manufacturing Applications (TIMA) research.

For project $A_{\text{without ATP}}$, the private rate of return is less than the private hurdle rate because of barriers to technology. As such, the private firm will not choose to invest in project A, although the social benefits from undertaking the project would be great. The vertical distance measured by the distance from i_{social} to the 45-degree line at i_{private} for project $A_{\text{without ATP}}$ is the spillover gap; it results from the additional value society would receive above what the private firm would receive if project $A_{\text{without ATP}}$ were undertaken. Project $A_{\text{without ATP}}$ is precisely the type of project in which the public should invest, namely one in which the private sector would not invest because of market failure and one from which society would greatly benefit.

Referring back to the interview information suggesting that the research projects would not have been undertaken or would have been undertaken at a reduced level or pace, we concluded that the TIMA research projects, once they have their support from ATP, have similar characteristics to those of Project $A_{\text{with ATP}}$ in Figure 9.1, yet without the ATP funding, they are similar to Project $A_{\text{without ATP}}$ in that each of the respondents viewed the expected private rate of return absent ATP funds to be less than his firm's private hurdle rate. Hence, these TIMA projects were valid candidates for ATP support in the Jaffe sense, and in the broad sense these are the types of projects that a public/private partnership competition policy could affect.

In Figure 9.2, we alternatively illustrate that reduction in risk in terms of a rightward shift in the distribution of the rate of return for the private firms. We illustrate the rate of return in Figure 9.2 to show our conceptualization of operational risk—risk is defined operationally below in terms of reducing the probability of a rate of return below the private hurdle rate. The rightward shift of the distribution, and the concept of reducing the probability of returns lower than acceptable to the private investors, applies equally well to the absolute level of net return (absolute return minus private investment) expected from the project.

For each distribution in Figure 9.2—without ATP funding (left distribution) and with ATP funding (right distribution)—the expected rates of return are shown.[7] As drawn, with ATP funding the expected private rate of return and the variance in the private rate of return from the research project will increase. One can generalize that this will always be the case.[8]

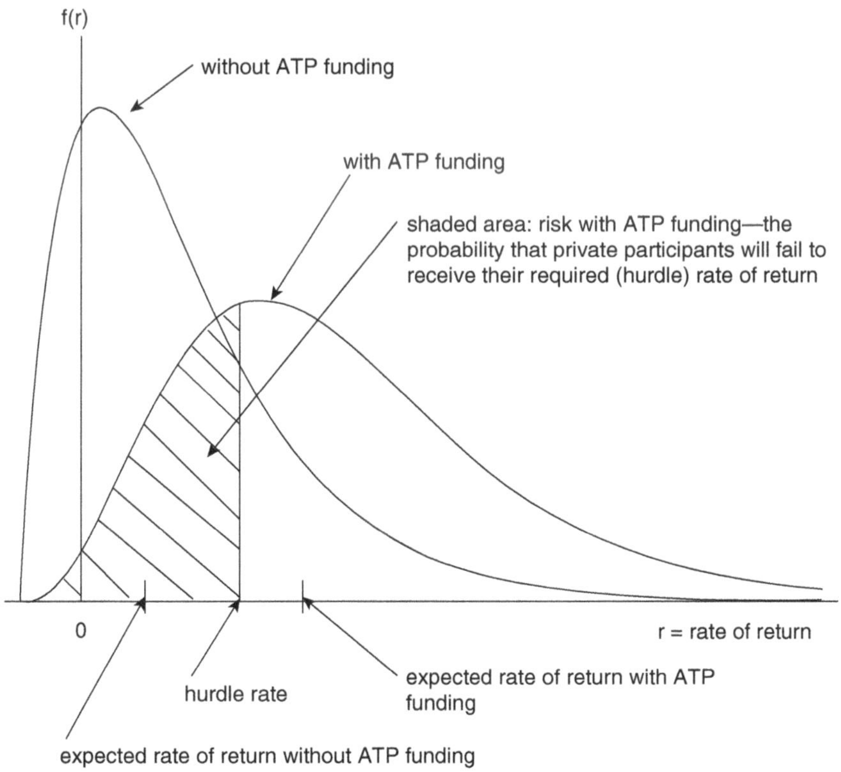

Figure 9.2 Private risk reduction resulting from Advanced Technology Program (ATP) funding.

Consider the left distribution—the distribution of the rate of return for the private firm without ATP funding. As drawn, the private hurdle rate is to the right of the expected rate of return without ATP funding, meaning that the private firm will not undertake this research because the firm will not receive its required rate of return. The risk of the project equals the area under this without-ATP distribution that is to the left of the private hurdle rate. For those used to thinking of the variance of the distribution as the measure of risk, the downside risk—which is the probability of a rate of return less than the hurdle rate—might seem unusual. Variance measures the possibility that outcomes can differ from the expected outcome, while the downside risk measures the probability of an outcome departing to the downside of the hurdle rate. Note that the technical risk and the market risk for the project are reflected in the variance of the distribution—the technical goals may exceed or fall short of expectations, and market acceptance of the project's technical outcomes could do the same. The downside risk refers to the outcomes that fall short of the hurdle rate.

Consider the right distribution—the distribution of the rate of return for the private firm with ATP funding. With ATP funding, the private firm will expect a return greater than its hurdle rate—the expected private rate of return with ATP funding is drawn to the right of the private hurdle rate.[9] While ATP funding will not itself increase the probability that the research will be successful, assuming hypothetically that it were undertaken absent ATP funding, it will reduce private risk by increasing the expected private rate of return because the expected rate of return will be based on a smaller private outlay.[10] Hence, ATP funding leverages the private firm's investment as illustrated by a greater expected return and a greater variance in the distribution as explained earlier.

The shaded area in Figure 9.2 is what we term *the downside risk of the project*—that is, it is the probability that the project will yield a rate of return less than the private hurdle rate even with ATP funding. Hence, the amount of downside risk with ATP funding is visually less than the downside risk associated with the research project in the without-ATP funding case.

Although we will conclude that ATP funding reduces risk, as defined operationally in terms of reducing the probability of a rate of return below the private hurdle rate, we emphasize that our analysis that follows is in no way wed to any particular measure of risk or any particular model of capital asset pricing with associated systematic and nonsystematic risk. Instead, our treatment encompasses any and all such models because the relevant risk, however it is perceived by private firms, is captured in the private hurdle rate, and the distributions of returns are otherwise represented by their expected values. In describing the effect of ATP funding on the distribution of private rates of return, we are describing an underlying reality that would be reflected in the private hurdle rate—as determined by some model—and in the expected value of the returns.

9.3. Application of the Spillover Evaluation Model

Based on our extensive interviews with contact individuals in the TIMA projects, we collected quantitative information on the variables in the first two columns of Table 9.2.

We relied on these project-specific data to calculate the critical values shown above in Figure 9.1. In particular, we calculate the expected social rate of return associated with the TIMA projects, i_{social}; the private rate of return without ATP funding, $i_{private}$; and the private rate of return with ATP funding, $i^*_{private}$. Given our understanding developed in the interviews along with the derived estimates for these critical values we concluded the following:

- The private rate of return to TIMA projects is less than the private hurdle rate; hence, these projects would not have been pursued absent ATP support, as initially inferred from interviews.
- There are spillover benefits to society associated with these TIMA projects as evidenced by the size of the spillover gap, $(i_{social} - i_{private})$.

For our calculation of the expected social rate of return from TIMA research projects, we calculated a lower bound for the expected social rate of return in two different ways. First, we calculated the expected social rate of return for each of the eight TIMA projects studied, and then we averaged across the rates; and second, we first averaged the eight observations of project-specific data for each relevant variable, and then estimated one expected social rate of return based on average data for each variable. Neither of these methods precisely characterizes the expected social rate of return to the TIMA research program, but each approximates what might be called an expected social rate of return for an average TIMA research project.

Table 9.2 Descriptive Statistics on Values Used in the Estimation of the Social Rate of Return (n = 8)

Variable	Definition	Mean	Standard Deviation
d	Duration of the ATP project	2.51 years	0.578
C	Total cost of the ATP project	$12,749,125	1.16×10^7
A	ATP funding	$6,421,875	5.48×10^6
r	Private hurdle rate	0.25	0.063
z	Duration of the extra period of development	1.03 years	0.619
F	Additional cost for the extra period of development	$1,644,784	1.35×10^6
T	Life of the commercialized technology	7.69 years	3.96
m	Proportion of producer surplus appropriated by the participants in their particular application of the generic technology	0.675	0.152
k	Proportion of total applications of the generic technology developed by the project addressed by the participants' applications	0.216	0.118
v	Proportion of producer surplus appropriated	0.135	0.071
L	Lower bound for expected annual private return to investing firm	$5,250,439	5.59×10^6
U	Upper bound for expected annual private return to the investing firm	$9,885,615	1.05×10^7

ATP, Advanced Technology Program.

The calculation of the expected social rates of return using each of our two methods relied on information collected during the telephone interviews and information published by ATP about each project. Interview respondents were assured that their individual responses would remain confidential; hence, only the average values of interview data and derived metrics are reported herein. The mean values and standard deviations for all of these variables are in the last two columns of Table 9.2.

Data related to project duration, d, total project cost, C, and the level of ATP funding, A, were reported in ATP's Project Brief for each project provided to us by ATP. Data on the next six variables—r, z, F, T, m, and k—were obtained through the interview process.

Respondents were asked the following question about their private hurdle rate, *r*:

> For projects like this one, what is your firm's hurdle rate or minimum expected rate of return over the life of the project?

Regarding the duration of the extra period of development, *z*, and the additional research cost, *F*, for the extra period of development, respondents were asked the following questions:

> Approximately, how soon after the project's completion will the technology be commercialized?
> Approximately, what is the level of the additional investment expenditures expected to be made by your firm during this period?

Once commercialized, the life of the technology, *T*, was determined from responses to the question:

> Approximately, what is the projected life of the technology being developed?

The variable *v* represents the proportion of producer surplus appropriated by the project team; larger (smaller) values of *v* are interpreted to mean that less (more) spillover of technical knowledge is anticipated and thus appropriability problems are less (more) critical. From an analytical perspective, this is a critical value for our analysis. We calculated *v* as the product of two pieces of information collected from the interviews. It equalled the product of the proportion (*m*) of total profits anticipated from the development and commercialization of the specific software being developed, and the proportion (*k*) of all possible applications of the generic technology addressed by the specific software being researched in the ATP project.[11] Thus *v*, the product of these two interview elements, is the total proportion of the value of the technology appropriable by the researching firms. Loosely speaking, and focusing on the mean values in Table 9.2, the funded firms expect to appropriate about 70% (*m*) of the market value of the software they are developing in the 20% (*k*) (by value) of the applications areas they are focusing on. Roughly, the funded firms expect to capture about 14% of the profit potential of the generic technology being developed, and 86% of the profit potential associated with the generic technology being developed will spill over to imitators. (The variables *L* and *U* are discussed later.)

Given the published data in the ATP-prepared project briefs, and interview data on each of the variables noted in Table 9.2, the first step in the calculation of the expected social rate of return was to calculate the variable L, the lower bound for the annual expected private return from a TIMA research project. L was a derived variable. It was derived for each of the eight projects using project-specific data as reported in Table 9.2. It was also calculated separately using the averaged values of each relevant variable, as discussed in our following exposition of the average TIMA project.

Equation (9.1) consists of three general terms. Each term represents the present value for a particular flow that is realized over a particular time period. The first term in the equation represents the present value of the negative cash flows that result to the firm from the cost of conducting the project, $(C - A)$, from its start to its expected completion, ($t = 0$ to d). The second term is the present value of the future negative cash flows from the additional cost, F, of taking the generic technology from the ATP project, at ($t = d$), and commercializing it, at ($t = d + z$). Finally, the third term is the present value of the expected net cash flows from the project, L, after it has been commercialized, at ($t = d + z$), over its estimated life, to ($t = d + z + T$). Note that the discount rate in equation (9.1) is the firm's hurdle rate, r. Therefore, the value for L that solved equation (9.1) was the value for which the private firm just earns its hurdle rate of return on the portion of the total investment that it must finance. The firm would not invest in the ATP project unless it expected at least L for the average annual private return so that its hurdle rate would exactly be met. Thus, L was a lower bound estimate.

$$
\begin{aligned}
&-\int_0^d \left(\frac{C-A}{d}\right) e^{-rt}dt - \int_d^{d+z}\left(\frac{F}{z}\right)e^{-rt}dt + \int_{d+z}^{d+z+T} Le^{-rt}dt = 0 \\
&\Rightarrow -\left(\frac{C-A}{d}\right)\left(\frac{-1}{r}\right)e^{-rt}\Big|_0^d - \left(\frac{F}{z}\right)\left(\frac{-1}{r}\right)e^{-rt}\Big|_d^{d+z} + (L)\left(\frac{-1}{r}\right)e^{-rt}\Big|_{d+z}^{d+z+T} = 0 \\
&\Rightarrow \left(\frac{C-A}{dr}\right)\left(e^{-rd}-1\right) + \left(\frac{F}{zr}\right)\left(e^{-r(d+z)} - e^{-rd}\right) - \left(\frac{L}{r}\right)\left(e^{-r(d+z+T)} - e^{-r(d+z)}\right) = 0
\end{aligned} \tag{9.1}
$$

Given specific values for the variables d, C, A, r, z, F, and T, equation (9.1) was solved for the unknown variable, L. To illustrate using the mean values for these seven variables from Table 9.2, equation (9.1) solves for L equal to \$3,883,680. Again, this derived value of L is the lower bound for the estimate of the expected annual private return using the mean values of the relevant variables in Table 9.2. Since we will illustrate the solution of the three equations that follow also using the mean values of the relevant variables in Table 9.2, we will refer for ease of exposition to the solutions as those for the average TIMA project.

The second step in the calculation of the estimated social rate of return was to calculate the variable U, the upper bound for the expected annual private return for each TIMA project. U was derived for each of the eight projects, by solving equation (9.2) using values for the variables in Table 9.2. Note that the first term in equation (9.2) is the total negative cash flow from the cost of conducting the research project, C. Thus, U was an upper bound for the annual private return because a return greater than U would imply that the firm would earn a rate of return in excess of its hurdle rate in the absence of ATP funding, and therefore ATP funding would not be required for the project. Note that all reference to ATP is purposively absent in equation (9.2) because we were calculating the expected annual private return that would result in the private firm just meeting its hurdle rate in the absence of ATP involvement.

$$\begin{aligned}
&-\int_0^d \left(\frac{C}{d}\right)e^{-rt}dt - \int_d^{d+z}\left(\frac{F}{z}\right)e^{-rt}dt + \int_{d+z}^{d+z+T} Ue^{-rt}dt = 0 \\
&\Rightarrow -\left(\frac{C}{d}\right)\left(\frac{-1}{r}\right)e^{-rt}\Big|_0^d - \left(\frac{F}{z}\right)\left(\frac{-1}{r}\right)e^{-rt}\Big|_d^{d+z} + (U)\left(\frac{-1}{r}\right)e^{-rt}\Big|_{d+z}^{d+z+T} = 0 \\
&\Rightarrow \left(\frac{C}{dr}\right)\left(e^{-rd}-1\right) + \left(\frac{F}{zr}\right)\left(e^{-r(d+z)}-e^{-rd}\right) - \left(\frac{U}{r}\right)\left(e^{-r(d+z+T)}-e^{-r(d+z)}\right) = 0
\end{aligned} \tag{9.2}$$

To illustrate the solution of equation (9.2), values for the variables d, C, r, z, F, and T are given. Thus, equation (9.2) solved for U. To illustrate using the mean values for these given variables in Table 9.2, the derived value of U is \$7,267,910. This value represented the upper bound of the annual private return for what we call the average TIMA project.

Using the mean values (other than L and U) in Table 9.2, an estimate of the average expected annual private return to the firm was calculated as $[(L + U)/2]$, the mean of the upper and lower bounds on the average expected annual private return for the average TIMA project.[12] The average expected annual private return to the participating firm or firms equals v times the average expected annual return that will be captured by all producers using the technology (producer surplus). Knowing the average expected annual private return is $[(L + U)/2)]$ and knowing the portion of producer surplus that is appropriable, v, then total producer surplus equals $[(L + U)/2v]$, and hence this value is a lower bound for the average expected annual social return. It was a lower bound because consumer surplus has not been measured.

The expected private rate of return without ATP funding was the solution to i in equation (9.3), given solution values for L and U from equations (9.1) and (9.2). The solution value of i in equation (9.3) represented the rate of return that just equates the present value of the expected annual private return to the firm to the present value of research and postresearch commercialization costs to the firm in the absence of ATP funding.

$$-\int_0^d \left(\frac{C}{d}\right) e^{-it}dt - \int_d^{d+z} \left(\frac{F}{z}\right) e^{-it}dt + \int_{d+z}^{d+z+T} \left(\frac{L+U}{2}\right) e^{-it}dt = 0$$
$$\Rightarrow \left(\frac{C}{di}\right)\left(e^{-id} - 1\right) + \left(\frac{F}{zi}\right)\left(e^{-i(d+z)} - e^{-id}\right) - \left(\frac{L+U}{2i}\right)\left(e^{-i(d+z+T)} - e^{-i(d+z)}\right) = 0 \qquad (9.3)$$

The expected private rate of return without ATP funding, i, was estimated for each of the eight TIMA projects by solving equation (9.3) given the values for d, C, z, F, T, L, and U. To illustrate the solution of equation (9.3) using the mean values of the relevant variables in Table 9.2 along with the values derived from equations (9.1) and (9.2) of L and U for the average project, the expected private rate of return for the average TIMA project was 0.20 or 20%.

Finally, the lower bound on the social rate of return is found by solving equation (9.4) for i, given values for the other variables. Note that equation (9.4) is identical to equation (9.3) with the exception that the average expected annual private return, $[(L + U)/2]$, is replaced with the lower bound for the average expected annual social return, $[(L + U)/2v]$.

$$-\int_0^d \left(\frac{C}{d}\right) e^{-it}dt - \int_d^{d+z} \left(\frac{F}{z}\right) e^{-it}dt + \int_{d+z}^{d+z+T} \left(\frac{L+U}{2v}\right) e^{-it}dt = 0$$
$$\Rightarrow \left(\frac{C}{di}\right)\left(e^{-id} - 1\right) + \left(\frac{F}{zi}\right)\left(e^{-i(d+z)} - e^{-id}\right) - \left(\frac{L+U}{2iv}\right)\left(e^{-i(d+z+T)} - e^{-i(d+z)}\right) = 0 \qquad (9.4)$$

As with each of equations (9.1) through (9.3), equation (9.4) was estimated for each of the eight TIMA projects. Again, to illustrate the solution of equation (9.4) using the average values of d, C, z, F, T, and v from Table 9.2 with the derived values of L and U for the average project, the expected social rate of return for the average TIMA project was derived to be at least 0.63 or 63%.

To summarize, we have extracted an expected social rate of return from our model. We first designed a set of questions that allowed us to gather the additional information, to supplement the information in the project briefs, needed to derive an estimate of the private benefit stream anticipated by the participants. Second, we gathered information revealing the benefits created by the project that would be captured by producers that are not participants in the project. We then had an estimate of the total producers' benefits from the project to compare with the project's costs. Thus, we used the private participant's expected benefits and the project's costs to compute the private expected rate of return. We used the total expected benefits, to all producers regardless of whether they were participants, to compare to the total costs for the project to compute the lower bound on the social rate of return. It was a lower bound because we had not measured the consumer surplus created by the project for the ultimate consumers of the final goods that were made using the project's technology.

To summarize the findings from our analysis for the average TIMA research project, based on the mean values in Table 9.2 (excepting those for L and U) and the derived values from equations (9.1) through (9.4), there are two important points to be made. First, the average expected private rate of return in the absence of ATP funding was 20%, clearly less than the average private hurdle rate of 25% since the estimate of the upper bound for the average cash flows would just allow the hurdle rate to be reached. Thus, in the absence of ATP funding the TIMA firms would not have undertaken this research, and in fact they expressed this fact explicitly as noted earlier. Second, the expected social rate of return associated with ATP's funding of the TIMA projects was at least 63%, and hence the projects are expected to be socially valuable.

As discussed earlier, we estimated equations (9.1) through (9.4) for each of the eight projects. Doing so, we derived eight values for L and eight values for U (the averages for these values are shown in Table 9.2 and are somewhat higher than the single values derived for the average project and reported in the text), and eight lower-bound estimates of the expected social rate of return—one set of rates for each TIMA project. For each of the eight projects, just as for the average project, the private rate of return without ATP funding of course fell short of the private hurdle rate, yet because of the spillovers the projects were all socially valuable. The average of the eight expected private rates of return absent ATP funding was 19% (0.19 with a standard deviation of 0.06); the average expected social rate of return was 72% (0.72 with a standard deviation of 0.22). These derived rates of return were observably close to our average project's private and social rates of return of 20% and 63%, respectively. Thus, we were comfortable using these figures to characterize the average TIMA project.

Our broad-brush approach to estimating the various rates of return does not try to build in more detail for the model than the data can support. Our methodology gave a reasonable fix on the average expected annual private returns to the ATP project participants because we could solve for what they must be expecting as lower and upper bounds, and then we averaged the two expectations. What we did not know with certainty was the multiplier to apply to the participants' average annual expected returns to estimate returns to all producers. Our multiplier was based on the interview responses that estimate the variables m and k, and hence v. We do not have a good estimate of the lag from $(d + z)$, when the ATP project's technology was commercialized, until the technology was imitated not only by producers copying the particular application of the generic technology but also by producers developing new applications of the generic technology. Thus, because there will be some lag before imitation and because there will be some development costs for producers using the generic technology, the net average expected annual earnings to all producers using the generic technology will be somewhat less than $[(L + U)/2v]$ unless diffusion is instantaneous and costless. For the particular information technology being studied, the diffusion might well be relatively quick and additional development costs may well be minimal; however, the possibility that it was not the case can be modeled by lowering our multiplier $(1/v)$ to reflect the fact that if diffusion is not rapid, then the ATP participants' average annual returns were actually a larger proportion of the average social returns than the proportion v. To estimate the sensitivity of our results to the speed of the diffusion of the technology, we varied v in the following experiment. If v is increased by 10%, the estimated social rate of return for the average TIMA project falls from 63% to 61%; if v is increased by 50%, the estimated social rate of return falls to 54%; and if v is increased by 100%, the estimated social rate of return falls to 47%. Thus, our conclusion that ATP's TIMA projects were socially valuable is robust with respect to the parameter v when it is increased to capture slower diffusion rates (and any development costs to be netted out) for the applications of the generic technology.[13]

It is important to re-emphasize that the profit potential of the generic technology being developed that was appropriated by the TIMA firms represents producer surplus. Our analysis did not attempt to capture consumer surplus, and of course consumer surplus would not be appropriated by the researching firms. Thus, when we referred to a lower-bound estimate of the expected social rate of return, we were explicitly acknowledging that there are social benefits of the consumer surplus generated from the TIMA research-based software, but we are just not capturing them. Thus, our social rate of return estimate for the average TIMA project was clearly a lower-bound estimate.

Finally, we explicitly note that our model in equations (9.1) through (9.4) is a continuous time model, as is appropriate. However, if the model is approximated as a discrete time model, we calculated that the estimated social rate of return for the average TIMA project is again 63%, given the rounding of times to the nearest whole year and given appropriate placement (a particular choice of beginning or end of years that results in replicating the solution found with the continuous model) of the discrete cash flows. As would be expected, however, in the discrete model, the estimation is not insensitive to the placement of the cash flows, and for that reason we present and solve the more accurate continuous time model.

9.4. Interpretation of the Findings

The TIMA case study can be summarized as follows. The TIMA projects under study would not have been undertaken in the absence of ATP's public funding. Because of technical and market risk and because of appropriability issues—key elements of market failure—firms perceived that their expected rate of return had they pursued this research in the absence of ATP's support would have been less than their required rate of return. ATP funding of these projects reduced risk to the private firms. And, as a result of ATP funding, these projects were being researched and the expected social rate of return from them is estimated to be at least 63%.

Of course, one cannot conclude that a social rate of return of at least 63% is good or bad, or better or worse than expected. However, one can compare our estimate of the lower bound of the social rate of return to the opportunity cost of public funds. Following the guidelines set forth by the Office of Management and Budget (OMB, 1992) to use a real discount of 7% for constant-dollar benefit-to-cost analyses of proposed investments and regulations, then clearly a nominal social rate of return of 63% implies a real rate above 7% and thus is socially worthwhile.[14]

With reference to Figure 9.1, Project A characterizes the average TIMA project with and without ATP funding. We estimated private rate of return absent ATP funding of 20% compared to a private hurdle rate of 25%, and an estimated lower bound on the social rate of return of 63%. Hence, the spillover gap associated with the average TIMA project is at least 43%.

In comparison to these estimates, we calculated the private rate of return with ATP funding to be 33%. This rate came from the solution to equation (9.5) for i, given the mean values of d, C, A, z, F, and T from Table 9.2 and the derived values of L and U from equations (9.1) and (9.2).

$$-\int_0^d \left(\frac{C-A}{d}\right) e^{-it}dt - \int_d^{d+z} \left(\frac{F}{z}\right) e^{-it}dt + \int_{d+z}^{d+z+T} \left(\frac{L+U}{2}\right) e^{-it}dt = 0$$
$$\Rightarrow \left(\frac{C-A}{di}\right)\left(e^{-id}-1\right) + \left(\frac{F}{zi}\right)\left(e^{-i(d+z)} - e^{-id}\right) - \left(\frac{L+U}{2i}\right)\left(e^{-i(d+z+T)} - e^{-i(d+z)}\right) = 0 \tag{9.5}$$

The estimated rate of return with ATP funding of 33% is perhaps reasonably close to the private hurdle rate of 25%.[15,16] Had ATP funding been greater (less), the estimated rate of return with ATP funding would have been greater (less). However, there was no way for ATP to have calculated the optimal level of funding for these TIMA projects unless, as part of the focused program proposal, all relevant data, including hurdle rates, could have been assessed.[17] In the absence of such information, which in practice would be difficult to obtain because of, if nothing else, self-serving reporting by proposers, the funding scheme that ATP had implemented appears to be close to optimal.[18]

In principle, a bidding mechanism could be used by ATP to ensure that the private participants in TIMA research projects earn just a normal rate of return. Such a bidding mechanism is what Scott (1998) refers to as a hurdle-lowering auction.

Motivated by the case analysis of the ATP-funded TIMA projects, a broad public/private partnership competition policy should address three broad questions:

- How can the public get the best private partner for each partnership?

- How can society determine the optimal level of public funding for the partnership, not too much yet enough to overcome the underinvestment resulting from market failure?
- How can society avoid the potential for opportunistic behavior to which both the government and the private sector's partners are exposed?

Next we list several premises that we argue should motivate the formulation of a public/private competition policy, along with the implementation implications of those premises.

> *Premise 1:* The private sector knows more than the government about the investment characteristics of the technology projects—or at least has the resources to make the best estimate of the streams of returns and the risk.
> *Implication 1:* A competition policy should include a mechanism for setting up a public/private partnership that provides the incentive for private parties to determine who is best suited to be the private partner in a public/private partnership.
> *Premise 2:* The government desires to overcome the project's underinvestment resulting from market failure and to do so at the least cost to the public.
> *Implication 2:* A competition policy should include a mechanism that gives the selected private partner for the public/private partnership an incentive to undertake the desired level of investment while providing a proportion of the project's funding that is consistent with a normal expected private rate of return given the appropriability and risk characteristics of the project.
> *Premise 3:* All parties related to the public/private partnership want to overcome the potential for opportunistic behavior by the other party.
> *Implication 3:* Policy should include a mechanism that provides both the public- and the private-sector partners an incentive to participate in the project in a way that maximizes the total value of the project's expected outcome rather than the value to the individual partner that could of course use opportunistic behavior to benefit at the expense of the overall project's results.

General characteristics of the mechanism design: What are the general characteristics of the optimal mechanism design for public/private partnerships that will achieve the desired incentives for the private sector to choose the best private partner, for the private partner(s) to carry out the desired amount of investment at the least cost to the public, and for avoiding opportunistic behavior by either the public or the private partner?

Consideration of these questions suggests that the optimal mechanism design for what we advocate and call a public/private competition policy would have the private parties use a contingent valuation method to bid for the right to be the private partner. In particular, the bidding could be a hybrid bidding mechanism that combines an up-front bid, a periodic payment bid, and finally a royalty bid: private firms would bid for the right to be the private partner in the public/private partnership project that the government would fund.[19] Alternatively, instead of bids being accepted directly from the firms that will be performing the R&D, private venture capital firms that would manage the public investments might bid for the rights to manage the projects.

As a simple example of how a bidding process would deliver the public funding to a public/private partnership such as any of the TIMA projects discussed herein, consider the following. Suppose that from society's perspective, an R&D investment project would cost $100 now and generate the expectation of $130 in 1 year and nothing thereafter. Suppose further that the threshold rate of return justifying public funding—society's hurdle rate—is 10%. Thus, the R&D project yields a social rate of return of 30%, which exceeds society's hurdle rate of 10%, and of course the net present value of [(130/(1.1)) – 100] is greater than zero. Suppose that from a private perspective the project costs $100 and, because of incomplete appropriation of returns, yields the expectation of just $105 in 1 year. Suppose further that given the private risk the private hurdle rate is 15%. Thus, the private sector would not undertake the project, which has an internal rate of return of 5%, which is less than the hurdle rate of 15%; and, of course, net present value is then negative.

In the context of the foregoing example, which conceptually describes each of the TIMA projects, the bidding process would work as follows. The government announces that it will buy (literally) the R&D project, paying the $100 investment cost.[20] The government then opens the bidding for the right to be the private partner in the public/private partnership. Private firms will bid the amount $\$X$ such that $\$X(1.15) = \$105$, implying that $\$X = \$91.30$. The cost to the public of the project would then be $8.70. With great uncertainty about the future returns, the use of royalty bidding rather than the up-front bidding can yield more to the government. Additionally, private firms with better capabilities for doing the project would be expected to bid higher than those firms that are less well suited to the project.

Suppose the ATP, or the government in general, wanted to use a public/private partnership to develop a project. It would announce that it would provide an up-front payment of $\$F$ to support the R&D investment project to be conducted by the winning bidder in an auction to determine the private partner(s) for the public/private partnership (and when there are several partners as in an RJV we are viewing the RJV as the partner). Furthermore, the government pledges to provide a periodic flow of funds $\$C$ throughout the project's life to support the flow costs of the R&D project. The fixed cost $\$F$, and the flow cost $\$C$, correspond to the typical abstraction of the structure of costs for R&D investment projects.[21] Bidders vie for the right to be the private partner in the project by submitting a three-part bid—first, a bid for how much the private firm will pay the government up front; second, a bid on the periodic flow payment during the life of the R&D project; and third, a bid on the royalty rate that it would pay the government on the innovation produced by the public/private partnership and licensed (perhaps exclusively) to the private partner.

As McAfee and McMillan (1987) made clear, in the context of the appropriate combinations of assumptions about the characteristics of the asset being auctioned and the participants in the auction and their beliefs about the value of the asset, there are nontrivial choices to be made about the exact nature of the auction. Apart from the usual choices for auctions in general, there would be choices specific to the new institutional use of auctions to determine the private partner for the public/private partnership. For example, institutional arrangements must be designed to ensure that the government's payments of $\$F$ and $\$C$ go solely for the purchase of R&D investments; the private partner's profits from the R&D investment project will come after the innovation is introduced. However, we do not develop full details of the ideal auction in different circumstances; instead, we simply posit the basic idea and observe that the three-part bidding mechanism proposed has the potential for leveraging public funding optimally. If ATP had undertaken such a hurdle-lowering auction ex ante, in terms of Figure 9.1, the expected rate of return to the TIMA projects with ATP funding (actually estimated to be 33% in our model) would not have exceeded the project participants' private hurdle rate (25%). Hence, society would have been better off in terms of a more optimal allocation of R&D resources.

Finally, two things must be emphasized. First, the high social rates of return estimated and reported for the TIMA projects are very conservative, lower-bound estimates because they do not include consumer surplus in the benefit stream. The profits that will be generated by the technology are obviously a small proper subset of the social benefits that the technology will generate, but the estimation method measures only the return in the form of profits to the innovator and to other producers of the technology. Second, one might be sceptical about the TIMA respondents' earnest belief that the projects would not have been undertaken, or at least would not have been undertaken to the same extent or at the same speed, without ATP funding. If the research would have occurred without public funding, the estimated upper bound and hence the average of the upper and lower bounds for expected private returns would be too low, and the actual lower bounds for the social rates of return would be even higher than we have estimated. Furthermore, the gap between the social and private rates of return would remain, although that would not in itself justify the public funding of the projects.

Notes

1. This case study was conducted by Link and Scott in 2000; this chapter draws directly from Link and Scott (1998a, 2001a) and the Link and Scott references therein.
2. http://www.atp.nist.gov/atp/focusprg.htm
3. The 10 projects, along with the sponsoring firms, were as follows: Model-Driven Application & Integration Components for MES, sponsored by Vitria Technology, Inc.; An Agent-Based Framework for Integrated Intelligent Planning–Execution, sponsored by IBM Manufacturing Solutions Unit; Advanced Process Control Framework Initiative, sponsored by Honeywell Technology Center; Solutions for MES-Adaptable Replicable Technology (SMART), sponsored by National Industrial Information Infrastructure Protocols Consortium; Virtual Reality Telecollaborative Integrated Manufacturing Environment (VRTIME), sponsored by Searle; Process Integration Using Model-Driven Engines, sponsored by Vitria Technology, Inc.; Agent-Enhanced Manufacturing System Initiative, sponsored by Advanced Micro Devices; EECOMS: Extended Enterprise Coalition for Integrated Collaborative Manufacturing Systems, sponsored by IBM Corporation, CIIMPLEX; Distributed Factory System Framework; sponsored by Consilium, Inc.; and ANTS Scheduling and Execution System, sponsored by Denab Robotics, Inc.
4. One respondent was the contact person for two projects. Vitria Technology, Inc. chose not to participate in this study without explanation. Vitria Technology was involved in two research projects.
5. Jaffe did not quantify what a large spillover is. Neither he nor we can define such a concept in the absence of a benchmark. Rather, we present an estimate of the size of the spillover gap that characterizes the TIMA projects.

6. See Link (1998, 1999) and Link and Scott (1998, 2001a) for a more detailed discussion of public/private partnerships and related public policy relative to this diagram.

7. Note that the expected rate of return does not necessarily correspond to the greatest frequency or probability density because the distribution of rates of return need not be symmetric.

8. To capture the idea of limited liability for investors, we bound their return below by zero. Thus, the rate of return can be quite negative when the return falls below the amount invested, but because the return is bounded below at zero, the rate of return is bounded below by (–100%).

The expected private rate of return with ATP support is: r = [return – (total project cost – ATP funding)]/[total project cost – ATP funding]. Let Z = (total project cost – ATP funding). Then, r = (return – Z)/Z = [(return/Z) – 1]. The variance of r is: [$(1/Z)^2$ Var(return)], and it is a general proposition that as ATP funding increases (and hence Z decreases) the variance in the private rate of return increases (since [$1/Z$] gets larger). It is also a general proposition that the expected private rate of return = E[(return)/Z) – 1] must increase for the same reason. Furthermore, neither the expected social rate of return nor the variance in the social rate of return changes at all. The social cost is the same, and the social return is the same.

9. ATP funding need not affect the firm's private hurdle rate; that rate is set by corporate policy in most cases. Conceivably, because the operational measure of risk falls, the hurdle rate might fall as well in the presence of ATP funding, and the simulative effect of ATP funding would still hold.

10. Also, in general, it is possible that, apart from the funding itself, the support and guidance of the ATP program will lower the probability of low returns.

11. Seven of the eight participants offered an estimate of m and k. For the eighth project, we imputed a value of m and k equal to the average of the reporting seven. Thus, eight observations were used for the calculation of each mean in Table 9.2.

12. We wanted an estimate of the average expected annual return, and the simple average of the lower and upper bounds is the natural measure. Alternatively, one could use both L and U and obtain an upper and lower bound on each solution to equations (9.3), (9.4), and (9.5). One could then think of their average as a better estimate. However, our procedure is more direct and more simply exposited.

13. We thank Jeanne Powell from ATP for raising this interesting issue.

14. Link and Scott (1998a) discuss the use of this guideline for National Institute of Standards and Technology (NIST) economic impact assessments.

15. Our estimation procedure did not force the closeness of these rates. Although greater ATP funding (an increase in A) has no effect on our estimate of U, it does of course require that our estimate of L and hence of [$(L + U)/2$] be lower. However, the initial private investment [$C – A$] falls as ATP funding increases, and that effect dominates causing our estimate of the private rate of return to increase as ATP funding increases. For example, if ATP funding for the average TIMA project were increased by 50%, our estimate of the private rate of return with ATP funding would have risen from 33% to 42%.

16. The expected private rate of return with ATP funding was calculated for each of the eight projects. The mean of these eight rates is 0.35 or 35%, with a standard deviation of 0.094.

17. Note that it is mathematically possible to solve equation (9.5) for the level of ATP funding, A^*, that would equate the estimated private rate of return with ATP funding to the hurdle rate of 25%. Such an exercise would lack policy relevance in the sense that the calculation has the benefit of hindsight and information collected after ATP funding was allocated; furthermore, we do not know the true weights for our estimated upper and lower bounds. We assume the equal weights of 0.5 for each, but although our basic conclusions about the relation between the social and private rates of return are not sensitive to that assumption, the calculation of A^* is dependent on whether the actual expected annual returns are closer to the upper or the lower bound. We believe it is safer for the ATP funding to be on the high side in any case. In a separate study (Hall, Link, and Scott 1998, 2003), we have found that greater ATP funding dramatically reduced the probability of termination of ATP information technology projects. The theoretical reason is clear; having more ATP funds shifts the distribution of the project to the right and lowers the probability that the private participants in the project, who are constantly monitoring the progress of the project, would want to terminate the project because the probability of an unacceptably low rate of return was too high.

18. The aforementioned findings should be interpreted in light of three particular limitations of this chapter. First, the conclusions presented here are based in part on interview data collected from a key participant in each TIMA project. As such, there is obviously some uncertainty built into the expressed estimates, although they are the best point estimates available. Perhaps more important, this interview information may be time dependent, meaning that it was collected at an early point in the progress of the research project. As such, the estimates presented herein are not definitive estimates to be expected at the completion of each project, but rather characterizations of the expected social benefits at a point in time early in the project. Second, we have referred to our lower-bound estimate of 63%—although this caveat refers to any of the calculated social rates of return—as a metric to characterize the average TIMA project. And third, our analyses have not specified any diffusion paths. We refrained from going beyond the limits of our data and did not attempt to model, much less quantify, when in time rivals will imitate the participants' applications of the generic technology, or when in time others will commercialize other aspects of the generic technology. Still, with these limitations in mind, it appears at a point early in the progress of the TIMA projects that they are socially worthwhile.

19. There is a large literature describing bidding mechanisms in great detail. McAfee and McMillan (1987) provide an excellent review, and they set out the general hybrid mechanism with the up-front bid as well as the royalty bid. Hansen (1985) and Samuelson (1986) provide analyses of the royalty bidding and bidding for the up-front fee and the royalty rate simultaneously.

20. Martin and Scott (2000) provide detailed discussion of the circumstances in which market failure and underinvestment would be expected to occur; the discussion is needed to inform the identification of projects that would be funded.

21. See Lee and Wilde (1980).

10

Concluding Statement

10.1. Four Alternative Methods

We compiled this book to provide an explanation and illustration of four alternative evaluation methods for the estimation of the social benefits of publicly financed R&D. We hope that it provides a useful starting point for understanding the intricacies of program evaluation and that it provides general templates for others to follow, although we emphasize that each evaluation will be unique.

We expect that some may prefer other methods than the evaluation methods that we set forth. Certainly, academics and practitioners have been researching program evaluation methods and applying evaluation methods for decades, and quite naturally they have become comfortable with their own evaluation paradigms. Often, evaluation methods are discipline specific; certainly scholars have been involved in the evaluation of public-sector programs in areas other than publicly financed R&D, and these other areas cover quite different programs, ranging, for example, from education to public administration.

Nevertheless, we believe that three of the program evaluation methods proffered herein follow directly from economic theory, rather than from discipline or agency convention, and are sound and broadly applicable (although illustrated herein with examples of R&D-based programs only). Economic theory underlies the traditional, counterfactual, and spillover evaluation methods that we have illustrated herein with appropriate case studies. We have illustrated a fourth, non–economics-model-based approach that provides useful complementary information for evaluation and also is an alternative when data are insufficient to use the economics-based approaches.

10.2. Economic Theory as the Guiding Criterion for Program Evaluation

We believe that economic theory should be the guiding criterion for any method of program evaluation because it focuses attention on the value and the opportunity costs of the programs. Without that focus, effective discipline in public expenditures of taxpayers' claims on resources is lacking. Figure 10.1 reprises the market-failure justification for public expenditures discussed in Chapter 2 and illustrated there with Figure 2.1. As explained in Chapter 2, the project with private and social rates of return denoted by point A is socially valuable because its social rate of return exceeds the social hurdle rate. Yet the market system's private sector will not undertake it because its private rate of return falls short of the private hurdle rate; hence, public support for this project that society would want undertaken is needed to have the project performed.

Without the grounding in an enumeration of social and private costs and benefits, the market failures that justify public support of the project cannot be identified. Projects with private and social rate of return outcomes like those denoted by point B might receive public support, yet the private sector would have done the project without such support,

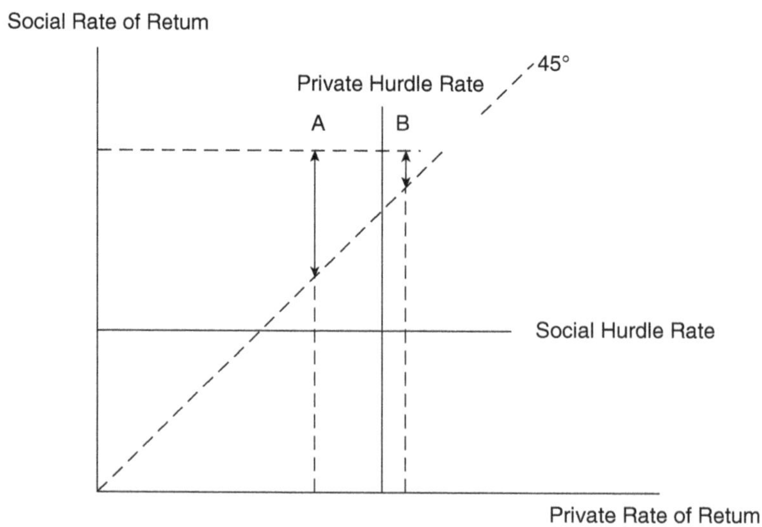

Figure 10.1 A reprise of the spillover gap.

since its private rate of return exceeds the private required rate of return. When the public supports a project, it uses claims on resources taken from the taxpayers, some of whom may be having difficulty paying for sufficient heating, shelter, food, or medical care. Taking their resources, or the resources of the taxpayers more generally, to have public funds for projects that the private sector would have undertaken efficiently without those resources would to many seem indefensible.[1] Evaluation methods grounded in the economic theory of value and opportunity costs—with careful attention to determining the social values and the social opportunity costs of resources—are needed to avoid indefensible uses of public funds, although as we explain in Chapter 4 and reprise briefly in the following discussion, the identification of benefits and costs is not a trivial matter. Yet an evaluation's attempt to identify them and carefully enumerate them in the context of an explanation of market failure is the key to meaningful decisions about the use of public funds.

10.2.1. Underinvestment and Inefficient Investment

In some cases like the one illustrated with the rates of return labeled as point B in Figure 10.1, market failure can occur, and the use of public funds can be justified. That will occur when the illustrated spillover gap has resulted not because of social returns that the private investors have not appropriated, but instead because the social costs of performing the research are less than the private costs. Thinking about the many economic impact studies done at the National Institute of Standards and Technology (NIST) will illustrate the point.

Tassey (2008) reported a summary of 19 NIST economic impact studies,[2] many of which used the counterfactual evaluation method that is appropriate for describing the benefits of publicly performed R&D for infrastructure technology:

> One of the key characteristics of measurement infratechnologies is their ubiquitous economic role. Every phase of the R&D stage and the subsequent manufacturing and commercialization (market development) stages of economic activity requires measurement infratechnologies and associated standards. Nineteen microeconomic studies conducted by NIST have shown the economic efficiency gains realized from the availability of this infrastructure to be substantial with an average benefit-cost ratio of 44:1. (p. 2)

With the microeconomic studies reviewed by Tassey, we begin with the expectation that NIST is involved when there is a market failure resulting in *(a)* an underinvestment in the activity or *(b)* an inefficient investment in it. That is, *(a)* the social value of the activity exceeds its social cost, yet the private value falls short of private costs; hence, the market does not provide socially desirable investment. Or *(b)* even if the market would make the investment, because of NIST's advantages as an honest broker—that is, its impartiality toward competing firms—and repository of experience and knowledge in the science of infrastructure technology, it would be able to do the investment more efficiently. That situation would, absent NIST's efficient provision of the infrastructure technology, represent a market failure, too.

The NIST economic impact studies reviewed by Tassey do not typically test the expectation *(a)*, although they do provide a great deal of indirect evidence to support it. Instead, the NIST economic impact studies have begun by assuming NIST is involved because of market failure and then the evaluation asks if NIST performed the work more efficiently than the private sector would have done. Thus, the existing studies do provide direct tests of case *(b)* market failure, but only indirect evidence about the case *(a)*.[3] In the context of such studies, there is usually a qualitative discussion of why there would be an underinvestment market failure absent NIST's involvement. But the benefit-to-cost ratios and related metrics simply ask whether having NIST do the project allowed the private sector to avoid costs that exceed what NIST and others, including those in the private sector, spent for the project. In all, then, the NIST economic impact studies reviewed by Tassey typically have a qualitative description of why in the particular case an underinvestment market failure is expected, and then the evaluation itself shows that NIST performed the project more efficiently than the private sector could have performed it, thus at the very least avoiding the market failure of inefficient investment (but probably also avoiding the underinvestment as well). Therefore, the public would not have wanted just to subsidize the private sector to have it perform the project.

In the course of doing an economic impact study in the usual way, we often come across information supporting the idea that the private sector would not have done the infrastructure technology investment—at least would not have done it to the extent that NIST did. But the key point is that the market failure story is the theoretical backdrop for why NIST is involved. Simply, NIST is, as supported by qualitative description of the project and its circumstances, doing something that the public wants done but that the private sector would not provide sufficiently or, as directly supported by the impact studies, would provide inefficiently. The economic impact studies using the counterfactual method do not test the first possibility of underinvestment but instead proceed to ask whether NIST performed the project more efficiently than the private sector could have done. As the data are collected, we often learn from industry that it simply would not have made the investment, but the NIST economic impact studies are not typically focused on demonstrating that point.

Observe that if one wanted to revisit the NIST economic impact studies and do more than a qualitative description of the possibility of underinvestment market failure, one would need to develop for the project absent any public support the private costs and private benefits and the social costs and social benefits, showing that social benefits exceeded social costs yet private benefits fell short of the private costs. Then, in that case, one could have either private performance with public support, or public performance. The NIST economic impact studies usually describe the likelihood of underinvestment market failure and the likelihood of inefficient (more costly) private performance, and then actually evaluate whether the public performance choice was an efficient one.

10.2.2. Caveats about Economics-Based Evaluations

There are important practical and theoretical controversies about benefit–cost ratios.[4] An important practical issue that makes it difficult to compare the benefit–cost ratios across individual studies is that some analysts are more conservative in tallying benefits and costs than others, even though all of the analyses use the same general formula for computing the benefit–cost ratio.[5] Also, some studies may be more or less successful at separating the counterfactual question, suitable for studying the benefits of publicly performed R&D, from the traditional question. Moreover, not all studies use the counterfactual approach; clearly the benefits measured with the traditional approach are not the same thing as the benefits measured with the counterfactual approach.[6] Another important theoretical issue, pointed out in Chapter 4, revolves around the use of market values to determine the opportunity costs and the values of the benefits for the investments. In the final section of this chapter, we conclude by observing that economics-based evaluations can help to discuss and resolve disagreements about the benefits created by investment programs and about the opportunity costs of the scarce resources used for the programs.

Despite these concerns, the economic impact studies Tassey (2008) summarized support the hypothesis that the evaluated NIST investments in infrastructure technology have generated social benefits that are many multiples of the costs. That conclusion points to another issue: the evaluations are for the projects selected for study and for those that responded in the necessary ways to the inquiries used to develop the evaluation. As we observed in Chapter 6, selection bias can result in performance for the selected sample that does not reflect the performance of the population of projects.

While the case studies herein do provide an illustration of each of the four evaluation methods discussed, the reader should realize that any application of a particular evaluation method will be unique. Thus, our case studies should not be interpreted as strict templates but rather general templates, for the four methods we illustrate, of the concepts to consider. Moreover, generalizations based on our case studies should be accompanied with caveats for concerns about sample size and about the ability to replicate evaluation results.

The sample size used in the four case studies was small by conventional econometric standards—ranging from 1 used in the estimation of consumer surplus in the case of SRM 2517a to several hundred in the case of publications and patents resulting from ATP's intramural research support. Thus, although some of our evaluation metrics, such as the expected number of publications from an ATP intramural project, can be viewed as point estimates grounded in a statistical model (and from the estimated model confidence intervals can be derived), other metrics such as the internal rate of return or benefit to cost ratio for a case study are not explicitly grounded in a statistical model because the numbers of observations are insufficient for formal statistical modeling. Clearly such metrics could be too large or too small; they combine the expert opinion of individuals with first-hand knowledge of the case to formulate a consensus opinion about benefits, but there is not a formal statistical analysis from which confidence intervals from statistical theory could be derived. Thus, the hypothesis testing associated with these metrics marshals a different type of evidence about the weight of expert opinion rather than about the statistical properties of large samples of data.

Turning to the issue of replicability, observe that evidence about economic theory is ideally based on revealed preferences, meaning that individual consumers and firms reveal their preferences through market transactions. However, as is generally the case, and certainly the case in the four case studies herein with exceptions such as actual counts of publications associated with a project, evaluation conclusions are drawn on the basis of expressed preferences, referring to what individuals express would have been their behavior in a hypothetical situation. This of course implies that the results from an evaluation exercise might not be replicable at another point in time because of interviewer or respondent biases.

To conclude our list of some of the more prominent practical caveats and special issues arising in evaluations, we point to Chapter 4's discussion of attribution. As we discussed in Chapter 4, when analysts develop the social benefits and social costs for evaluations, they should be aware that typically publicly financed R&D projects are actually public/private partnerships with opportunity costs incurred by both the public and the private sectors. Thus, our evaluation analyses, regardless of the evaluation method used, recognize that in many cases the measured benefits attributable to the publicly funded project are the result of a public/private partnership for which there are contributions from the private sector as well as the public sector. For example, in the TIMA case and in the Baldrige National Quality Award case, public funds are quite prominently combined with contributions to cost by the private sector. In such cases, we sum the private and public expenditures to have the total social costs. The social benefits are then compared to the social costs.[7] Alternatively, in those cases where the evaluated project has added incrementally to extant knowledge—such as when public R&D investment develops an improved SRM—the evaluation must develop the incremental increase in benefits created by the public's investment, weighing only those incremental benefits against the investment costs.

10.3. Evaluations as the Foundation of Informed Policy Debates

The evaluation methods that we have explained, discussed, and illustrated with case studies focus debate about public choice in the allocation of scarce economic resources on the value created and the opportunity costs of the resources used to obtain that value. Recognition of the many caveats discussed in the preceding section instills the awareness that although evaluation methods will set out information necessary for intelligent decisions about the use of public funds, the methods provide the basis for informed debates about policy rather than necessarily settling all debate. In the context of policy debates, the methods can be used to identify—or at least clarify disagreements about—the social and private values and the social and private costs of applying scarce economic resources to a publicly funded project. The economics-based evaluation methods not only allow the identification of situations where there is a market failure to provide socially valuable investments. Additionally, the evaluation methods can identify cases of government failure—situations where public funds have been or would be applied inappropriately or ineffectively.

Notes

1. See Chapter 3's discussion of the need for evaluations to address accountability to the taxpayers.

2. These are available at http://www.nist.gov/director/planning/strategicplanning.htm

3. Our case study in Chapter 9 that uses the spillover method does provide direct evidence of market failure in the sense of case *(a)*. The case study using the spillover method is an evaluation of privately performed but publicly supported investment, rather than the publicly performed investments that the economic impact studies at NIST address.

4. See, for example, the discussion and references at the end of Chapter 4.

5. The number of economic impact studies exceeds the number of analyst teams that have done the studies. Thus, a statistical analysis of the benefit–cost ratios for the entire collection of economic impact studies would be expected to show significant "analyst-team effects," *ceteris paribus*.

6. An exception is explained in endnote 7 of Chapter 4.

7. Although we believe the best approach is to keep track of the private as well as public contributions and total the two to have the social costs to compare with the social benefits, another approach would consider just the public costs and then simply select an attribution percentage on the basis of ad hoc conservativeness. One example of such an approach was employed by the National Research Council (2001, p. 169), where, as part of their evaluation of a large number of Department of Energy (DOE) programs, they stated, "Attribution of these benefits to DOE is difficult to determine. However, since DOE did play an influential role in developing the technology, the committee *believes* [emphasis added] that it is *reasonable* [emphasis added] to attribute one-half of [the benefits] to it."

References

Act of July 11, 1890, U.S. Statutes at Large 26 (1890): 254.
Act of July 28, 1866, Public Law 36-183, *U.S. Statutes at Large* 14 (1866): 339.
Act of March 3, 1901, Public Law 56-117, *U.S. Statutes at Large* 31 (1901): 1449.
Act of February 14, 1903, Public Law 57-87, U.S. Statutes at Large 32 (1903): 825.
Act of July 21, 1950, Public Law 81-506, *U.S. Statutes at Large* 64 (1950): 370.
Act of June 20, 1956, Public Law 84-452, *U.S. Statutes at Large* 70 (1956): 319.
American Technology Preeminence Act of 1991, Public Law 102-245, *U.S. Statutes at Large* 106 (1991): 18.
Arrow, K. J. 1962. Economic welfare and the allocation of resources for invention. In *The rate and direction of inventive activity*, 609-626, Princeton, NJ: Princeton University Press.
Arrow, K. J. 1963. *Social choice and individual values*. 2nd ed. New Haven, CT: Yale University Press.
Audretsch, D. B. 1995. *Innovation and industry evolution*. Cambridge, MA: MIT Press.
Baldwin, W. L., and J. T. Scott. 1987. *Market structure and technological change*. London: Harwood Academic Publishers.
Banker, R. D, I. Khosla, and K. K. Sinha. 1998. Quality and competition. *Management Science* 44: 1179–1192.
Bator, F. M. 1958. The anatomy of market failure. *Quarterly Journal of Economics* 72: 351–379.
Black, S. E., and L. M. Lynch. 1996a. Human capital investments and productivity. *American Economic Review* 86: 263–267.
Black, S. E., and L. M. Lynch. 1996b. How to compete: The impact of workplace practices and information technology on productivity. *Review of Economics and Statistics* 83: 434–445.
Blodgett, N. 1999. Service organizations increasingly adopt Baldrige model. *Quality Progress* 5: 74–76.

Chief Financial Officers Act of 1990, Public Law 101-576, *U.S. Statutes at Large* 105 (1990): 2838.

Cole, F. J., and N. B. Eales. 1917. The history of comparative anatomy. *Science Progress* 11: 578–596.

Collins, E. 1997. *Performance reporting in federal management reform.* National Science Foundation special report, Washington, DC: National Science Foundation.

Council on Competitiveness. 1995. *Building on Baldrige: American quality for the 21st century.* Washington, DC: Council on Competitiveness.

Council of Economic Advisers. 1994. *Economic report of the President.* Washington, DC: U. S. Government Printing Office.

Council of Economic Advisers. 2000. *Economic report of the President.* Washington, DC: U. S. Government Printing Office.

Council of Economic Advisers. 2001. *Economic report of the President.* Washington, DC: U. S. Government Printing Office.

Cozzens, S. E. 1995. *Assessment of fundamental science programs in the context of the Government Performance and Results Act (GPRA).* Washington, DC: Critical Technology Institute.

David, P. A. 1987. Some new standards for the economics of standardization in the information age. In *Economic policy and technological performance,* ed. P. Dasgupta and P. Stoneman, 206-234, Cambridge, England: Cambridge University Press.

Easton, G. S., and S. L. Jarrell. 1998. The effects of total quality management on corporate performance: An empirical investigation. *The Journal of Business* 71: 253–307.

Executive Office of the President. 1990. *U.S. technology policy.* Washington, DC: Office of Science and Technology Policy.

Executive Office of the President. 1994. *Science in the national interest.* Washington, DC: Office of Science and Technology Policy.

Executive Office of the President. 1998. *Science and technology: Shaping the twenty–first century.* Washington, DC: Office of Science and Technology Policy.

Executive Office of the President. 2003. President's 2004 Budget. http://www.gpoaccess.gov/usbudget/fy04/pdf/budget/performance.pdf.

Federal Financial Management Improvement Act of 1996, Public Law 104-208, *U.S. Statutes at Large* 110 (1996): 3009.

Frank, R. H. 2000. Why is cost-benefit analysis so controversial? *Journal of Legal Studies* 29: 913–930.

Gilbert, S. L., and W. C. Swann. 2001. *Acetylene* $^{12}C_2H_2$ *absorption reference for 1510 nm to 1540 nm wavelength calibration*: SRM 2517a. NIST Special Publication 260–133. Gaithersburg, MD: National Institute of Science and Technology.

General Accounting Office. 1985. *Managing the cost of government—building an effective financial management structure.* Washington, DC: U. S. General Accounting Office.

General Accounting Office. 1991. *Management practices: U.S. companies improve performance through quality efforts.* Washington, DC: U. S. General Accounting Office.

General Accounting Office. 2004. *Performance budgeting: Observations on the use of OMB's Program Assessment Rating Tool for the fiscal year 2004 budget.* Washington, DC: U. S. General Accounting Office.

Government Management Reform Act of 1994, Public Law 103-356, *U.S. Statutes at Large* 110 (1994): 3009.

Government Performance and Results Act of 1993, Public Law 103-62, *U.S. Statutes at Large* 107 (1993): 285.

Greene, W. H. 2002a. *LIMDEP version 8.0 within Nlogit 3.0, version 3*. Plainview, NY: Econometric Software, Inc.

Greene, W. H. 2002b. *LIMDEP version 8.0. Econometric modeling guide*. Vol. 2. Plainview, NY: Econometric Software, Inc.

Greene, W. H. 2003. *Econometric analysis*. 5th ed. Upper Saddle River, NJ: Prentice Hall.

Griliches, Z. 1958. Research costs and social returns: Hybrid corn and related innovations. *Journal of Political Economy* 66: 419–431.

Hall, B. H., A. N. Link, and J. T. Scott. 1998. *Universities as partners in ATP–funded research projects*. Final report for the Advanced Technology Program, Gaithersburg, MD: National Institute of Standards and Technology.

Hall, B. H., A. N. Link, and J. T. Scott. 2001. Barriers inhibiting industry from partnering with universities: Evidence from the advanced technology program. *Journal of Technology Transfer* 26: 87–98.

Hall, B. H., A. N. Link, and J. T. Scott. 2003. Universities as research partners. *Review of Economics and Statistics* 85: 485–491.

Hansen, R. G. 1985. Auctions with contingent payments. *American Economic Review* 74: 862–865.

Hao, K. Y., and A. B. Jaffe. 1993. Effect of liquidity on firms' R&D spending. *Economics of Innovation and New Technology* 2: 275–282.

Helton, B. R. 1995. The baldie play. *Quality Progress* 28: 43–45.

Hendricks, K. B., and V. R. Singhal. 2001. The long-run stock price performance of firms with effective TQM programs. *Management Science* 47: 359–368.

Huselid, M. A., and B. E. Becker. 1996. High performance work systems and firm performance: Cross–sectional versus panel results. *Industrial and Labor Relations Review* 54: 737–774.

Jaffe, A. B. 1998. The importance of 'spillovers' in the policy mission of the ATP. *Journal of Technology Transfer* 23: 11–19.

Ichniowski, C., K. Shaw, and G. Prennushi. 1995. *The effects of human resource management practices on productivity*. NBER working paper no. 5333. Cambridge, MA: National Bureau of Economic Research.

Lau, R. S. M. 1996. A survey of competitiveness and manufacturing practices. *South Dakota Business Review* 54: 1–9.

Lawler, E. E., S. A. Mohrman, and G. E. Ledford, Jr. 1995. *Creating high performance organizations: Practices and results of employee involvement and total quality management in Fortune 1000 companies*. San Francisco: Jossey–Bass.

Le Grand, J. 1991. The theory of government failure. *British Journal of Political Science* 21: 423–442.

Lee, T., and L. L. Wilde. 1980. Market structure and innovation: A reformulation. *Quarterly Journal of Economics* 94: 429–436.

Leyden, D. P., and A. N. Link. 1999. Federal laboratories as research partners. *International Journal of Industrial Organization* 17: 572–592.

Link, A. N. 1996a. *Economic impact assessment guidelines for conducting and interpreting assessment studies*. NIST planning report 96–1. Gaithersburg, MD: National Institute of Standards and Technology.

Link, A. N. 1996b. *Evaluating public sector research and development*. New York: Praeger Publishers.

Link, A. N. 1998. *Public/private partnerships as a tool to support industrial R&D: Experiences in the United States*. Report prepared for the working group on technology and innovation policy, Division of Science and Technology, Paris: Organization for Economic Cooperation and Development.

Link, A. N. 1999. Public/private partnerships in the United States. *Industry and Innovation* 6: 191–216.

Link, A. N. 2006. *Public/private partnerships: Innovation strategies and policy alternatives*. New York: Springer.

Link, A. N., and J. T. Scott. 1998a. *Overcoming market failure: A case study of the ATP focused program on technologies for the integration of manufacturing applications TIMA*. Final report prepared for the Advanced Technology Program, Gaithersburg, MD: National Institute of Standards and Technology.

Link, A. N., and J. T. Scott. 1998b. *Public accountability: Evaluating technology–based institutions*. Norwell, MA: Kluwer Academic Publishers.

Link, A. N., and J. T. Scott. 2000. Estimates of the social returns to small business innovation research projects. In *The small business innovation research program*, ed. C. Wessner, 275–290, Washington, DC: National Academy Press.

Link, A. N., and J. T. Scott. 2001a. Public/private partnerships: Stimulating competition in a dynamic market. *International Journal of Industrial Organization* 19: 763–794.

Link, A. N., and J. T. Scott. 2001b. *Economic evaluation of the Baldrige National Quality Program*. NIST planning report 01–3. Gaithersburg, MD: National Institute of Standards and Technology.

Link, A. N., and J. T. Scott. 2004a. *Evaluation of ATP's Intramural Research Awards Program*. NIST report 04–866. Gaithersburg, MD: National Institute of Standards and Technology.

Link, A. N., and J. T. Scott. 2004b. *The role of public research institutions in a national innovation system: An economic perspective*. Final report, New York: The World Bank.

Link, A. N., and J. T. Scott. 2005a. *Evaluating public research institutions: The U.S. Advanced Technology Program's intramural research initiative*. London: Routledge.

Link, A. N., and J. T. Scott. 2005b. Evaluating public sector R&D programs: The Advanced Technology Program's investment in wavelength references for optical fiber communications. *Journal of Technology Transfer* 30: 241–251.

Link, A. N., and J. T. Scott. 2006. An economic evaluation of the Baldrige National Quality Program. *Economics of Innovation and New Technology* 15: 83–100.

Link, A. N., and J. T. Scott. 2009. Evaluating public-sector intramural research programmes: The case of the U.S. Advanced Technology Program's intramural research initiative. *International Journal of Public Policy* 4: 532–548.

Link, A. N., and D. S. Siegel. 2003. *Technological change and economic performance*. London: Routledge.

Link, A. N., and G. Tassey. 1987. *Strategies for technology-based competition: Meeting the new global challenge*. Lexington, MA: D. C. Heath.

McAfee, R. P., and J. McMillan. 1987. Auctions and bidding. *Journal of Economic Literature* 25: 699–738.

Malcolm Baldrige National Quality Improvement Act of 1987, Public Law 100-107, *U.S. Statutes at Large* 101 (1987): 72.

Mansfield, E., J. Rapoport, A. Romeo, S. Wagner, and G. Beardsley. 1977. Social and private rates of return from industrial innovations. *Quarterly Journal of Economics* 91: 221–240.

Martin, S., and J. T. Scott. 2000. The nature of innovation market failure and the design of public support for private innovation. *Research Policy* 29: 437–447.

Narin, F. 1976. *Evaluative bibliometrics: The use of publication and citation analysis in the evaluation of scientific activity*. Washington, DC: National Science Foundation.

Narin, F., D. Olivastro, and K. A. Stevens. 1994. Bibliometrics/theory, practice, and problems. *Evaluation Review* 18: 65–76.

National Institute of Standards and Technology. 1996. Study finds 'quality stocks' yield big payoff. mimeograph, National Quality Office, Gaithersburg, MD: National Institute of Standards and Technology.

National Institute of Standards and Technology. 1997a. Results of Baldrige winners' common stock comparison, third NIST stock investment study. mimeograph, National Quality Office, Gaithersburg, MD: National Institute of Standards and Technology.

National Institute of Standards and Technology. 1997b. *Malcolm Baldrige National Quality Award: 1997 Criteria for Performance Excellence*. Gaithersburg, MD: National Institute of Standards and Technology.

National Institute of Standards and Technology. 1998. *Ten years of business excellence for America*. Gaithersburg, MD: National Institute of Standards and Technology.

National Institute of Standards and Technology. 2000. 'Baldrige Index' outperforms S&P 500 by almost 5 to 1. mimeograph, National Quality Office, Gaithersburg, MD: National Instiute of Standards and Technology.

National Research Council. 2001. *Energy research at DOE: Was it worth it?* Washington, DC: National Academy Press.

National Science Board. 1972. *Science indicators, 1972*. Washington, DC: National Science Foundation.

National Science Board. 2004. *Science indicators, 2004*. Washington, DC: National Science Foundation.

National Science and Technology Council. 1996. *Assessing fundamental science*. NSTC monograph. Washington, DC: National Science and Technology Council.

Office of Management and Budget. 1992. *Circular no. A–94: Guidelines and discount rates for benefit–cost analysis of federal programs*. Washington, DC: Government Printing Office.

Office of Management and Budget. 1996. *Economic analysis of federal regulations under executive Order 12866*. Washington, DC: Government Printing Office.

Office of Management and Budget. 2003. *Circular no. A–4: Regulatory analysis*. Washington, DC: Government Printing Office.

Office of Management and Budget. 2006. *Circular no. A-11: Preparation, submission and execution of the budget*. Washington, DC: Government Printing Office.

Omnibus Trade and Competitiveness Act of 1988, Public Law 100-418, *U.S. Statutes at Large* 102 (1988): 1107.

Orszag, P. R. 2009. *Memorandum for the heads of executive departments and agencies*. Washington, DC: Office of Management and Budget, Executive Office of the President, October 7.

Ruegg, R., and I. Feller. 2003. *A toolkit for evaluating public R&D investment models, methods, and findings from ATP's first decade*. Gaithersburg, MD: National Institute of Standards and Technology.

Samuelson, W. F. 1986. Bidding for contracts. *Management Science* 32: 1533–1550.

Scherer, F. M. 1979. The welfare economics of product variety: An application to the ready-to-eat cereals industry. *Journal of Industrial Economics* 28: 113–134.

Scott, J. T. 1998. Financing and leveraging public/private partnerships: The hurdle–lowering auction. *STI Science, Technology, Industry Review* 23: 67–84.

Scott, J. T. 2000. The directions for technological change: Alternative economic majorities and opportunity costs. *Review of Industrial Organization* 17: 1–16.

Scott, J. T. 2009a. Competition in research and development: A theory for contradictory predictions. *Review of Industrial Organization* 34: 153–171.

Scott, J. T. 2009b. Cost-benefit analysis for global public-private partnerships: An evaluation of the desirability of intergovernmental organizations entering into public-private partnerships. *Journal of Technology Transfer* 34: 525–599.

Sen, A. 2000. The discipline of cost-benefit analysis. *Journal of Legal Studies* 29: 931–952.

Sidgwick, H. 1885. *The scope and method of economic science*. London: Macmillan and Co.

StataCorp. 2001. *Stata statistical software, release 7.0*. College Station, TX: Stata Corporation.

Stalebrink, O. J. 2009. National performance mandates and intergovernmental collaboration. *The American Review of Public Administration* 39: 619-639.

Tassey, G. 1992. *Technology infrastructure and competitive position*. Norwell, MA: Kluwer Academic Publishers.

Tassey, G. 1997. *The economics of R&D policy*. Westport, CN: Quorum Books.

Tassey, G. 2003. *Methods for assessing the economic impacts of government R&D*. NIST planning report 03–01. Gaithersburg, MD: National Institute of Standards and Technology.

Tassey, G. 2005. Underinvestment in public good technologies. *Journal of Technology Transfer* 30: 89–113.

Tassey, G. 2008. *The roles and economic impacts of technology infrastructure*. mimeograph, Gaithersburg, MD: National Institute of Standards and Technology.

Teece, D. J. 1980. Economies of scope and the scope of the enterprise. *Journal of Economic Behavior and Organization* 1: 223–247.

Townsend, P., and J. Gebhardt. 1996. The importance of the Baldrige to the US economy. *Journal for Quality and Participation* July/August: 6–13.

U. S. Census Bureau. 1997. *1997 economic census, 1987 SIC basis, sales, receipts or shipments in 1997*. Washington, DC: U. S. Government Printing Office.

Willig, R. D. 1976. Consumer's surplus without apology. *American Economic Review* 66: 589–597.

Winston, C. 2006. *Government failure versus market failure*. Washington, DC: Brookings Institution Press.

Wisner, J. D., and S. G. Eakins. 1994. A performance assessment of the U. S. Baldrige quality award winners. *International Journal of Quality and Reliability Management* 11: 8–25.

Wolf, C., Jr. 1979. A theory of nonmarket failure. *Journal of Law and Economics* 22: 107–139.

Index

Page numbers followed by *f*, *t*, or *n* indicate figures, tables, or notes, respectively.

www.ingramcontent.com/pod-product-compliance
Ingram Content Group Australia Pty Ltd
76 Discovery Rd, Dandenong South VIC 3175, AU
AUHW021803050925
416304AU00004B/41

9 780199 729685